The Dictionary of Geographical Literacy

The complete geography reference

Kieran O'Mahony

EduCare Press Seattle, WA.

PUBLISHED BY
EduCare Press
PO Box 31511, Seattle, Washington 98103

Library of Congress Cataloging-in-Publication Data: 93-70539
O'Mahony, Timothy Kieran.
Dictionary of Geographical Literacy / Kieran O'Mahony

ISBN: 0-944638-08-2 $19.95

Printed and bound in the United States of America.

1 2 3 4 5 6 7 8 9
Cover design:
EduCare Press, Seattle

Contents

Other titles from *EduCare Press:*

Geographical Literacy
What we should know about geography
Kieran O'Mahony

Geography and Education
Through the Souls of Our Feet
Kieran O'Mahony

The Valiant Captains
Epics of the Sea - Historical
Sheldon A. Jacobson

The Man Who Moved the World
Archimedes - Historical
Sheldon A. Jacobson

Fleet Surgeon to Pharaoh
Ancient Egypt - Historical
Sheldon A. Jacobson

To The Woods and Waters Wild
Short Stories
Ross Brinn

Acknowledgments

The author wishes to acknowledge the many individuals and organizations who kindly cooperated in the production of this book, including Bernadette who contributed editorial assistance and read the proofs.

The author is greatly indebted to his colleague Paul Harris, F.R.G.S., for providing vivid photographs of various geographical phenomena and for his unique first-hand account of Russia Far East. I warmly thank John O'Leary and John Hartnett for their pertinent comments on initial drafts and for supplying many valuable suggestions concerning the layout of the book.

The following people offered expert advice and helpful comments relating to the selection and presentation of geographical entries: Professor Michael Meehan, Physical Geography, Professor Charlie O'Connell, Geomorphology, Noel Locke, Social Geography, Fernando Figueredo, South American Geography, Anthony Doherty, Human and Economic Geography, and Dr. Barry Brunt, Geography of the European Community.

The author particularly wishes to acknowledge the longtime friendship, support and encouragement offered by John K. Barry, Ph.D. and Jack Damron, Ph.D.

The author must also gratefully remember the late Dr. E. Fahy who inspired his students to great geographic quests.

Educational reforms are so intimately connected with politics, with problems of race, nationality, language and religious and social ideals, that they cease to be of narrow professional significance and have become a matter of general interest as the main problem of democratic government.

Nicholas Hans
Comparative Education

Preface

The degree of popular euphoria over the emergence of geography in recent years has been tempered by the conspicuous and widespread confusion that exists in relation to its nature and definition. It is common for people to think that a knowledge of geography refers simply to mountains, rivers and capital cities. This may have been true in the past, but modern geography is a much wider and more interesting subject; a subject that has captured the minds and challenged the imaginations of thousands of students and teachers across the nation.

The object of study in geography is the earth, and more precisely, the earth as the home of man. Consequently, the relationship between man and his environment is a recurring theme in this domain. Physical geography deals with the physical fabric of the earth and human and economic geography deals with the processes and activities of society in relation to the earth. It has become an important field of study in schools and universities today, since the global realization that geography influences almost everything we do in our world.

Literacy, as applied to geography, involves a fresh new layer on the educational itinerary. It presumes that one already has the ability to read and write and to communicate verbally. A dictionary of geographic literacy enhances this learning by providing a quick, and easy reference of core geographic knowledge. This cannot be an encyclopedic presentation, since a comprehensive treatment of all geographic phenomena and concepts would be a multi-volume, compendious production and, thus, far too cumbersome to nourish the present deficiency. Nor should this be seen as a limitation to acquiring profound principles of geography. On the contrary, its succinctness, its clarity and its illustrations provide a pragmatic solution for society's ubiquitous lack of geographical dexterity and proposes a solid foundation for the acquisition of meaningful geographic knowledge and skills.

Any work of this substance must be, by its very nature, subjective and, thereupon, ample concern for debate in relation to the items that were selected and those that were excluded. However, the underlying philosophy for the selection of geographical concepts and ideas was predicated by the need to furnish broad, pertinent core-information for a general and sanguine understanding of the subject. The author believes that the essence and application of geography enriches our lives and a literacy in its methods will make each individual more informed, more interesting and more responsible.

Kieran O'Mahony
Seattle, 1993

Aberdeen

A seaport on the east coast of northern Scotland. It is a main port and distribution center for products of the region. In the 1970's the discovery of natural gas and oil in the North Sea heralded a new wave of prosperity and development for Scotland and in particular for Aberdeen because of its strategic location. New energy-related jobs supplement those traditional livelihoods in agriculture and fishing.

Abominable Snowman

Also known as the "Yeti"; it is a large, hairy manlike creature reputed to exist in the Himalayan highlands. Tracks have been reported high on the slopes of some of the world's highest peaks.

Aborigine

The first inhabitants or plants of a region. A native. In Australia the native people are called Aborigines. Their ancestors heralded from Southeast Asia about 40,000 years ago. When European settlers arrived in Australia in the late eighteenth century the Aborigines were displaced and many died of diseases introduced by the new settlers. Today the Aborigines are trying to regain much of their ancestral lands and preserve their ancient culture in a modern and increasingly tolerant Australia.

Abrasion

A process of wearing down or rubbing away by means of friction. Wind and water can erode, or wear-away by rubbing or scraping against rocks. Often, unusual rock formations result. In deserts the wind will pick up grains of sand and blow them aloft, up to about one meter in height above the ground. Rock that is polished and sculpted in this manner will give the landscape a peculiar personality. These formations are called ventifacts and include yardangs and zeugens. In a similar fashion, water can abrade or wear-away a river bed, its sides and banks, as well as any rocks and pebbles along its course. Pebbles that have been abraded by hydraulic action are usually rounded and smooth.

Absolute Magnitude

The amount of light a star sends out is known as its absolute magnitude and by studying the brightness of the star we can estimate its distance from us.

Abyssal plain

A flat area of sediment on the ocean floor, between the continental shelf and the mid oceanic ridge. It is profoundly deep.

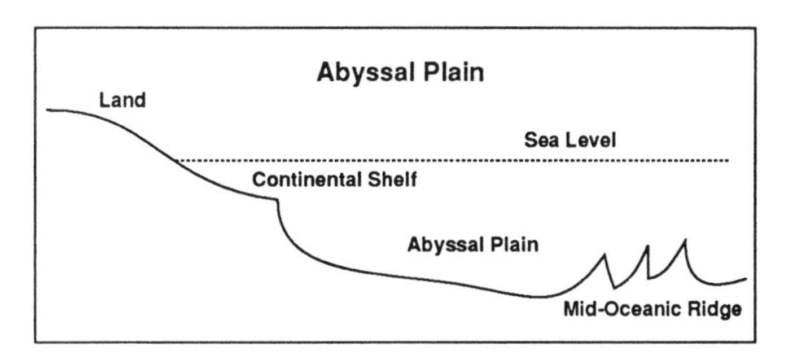

Fig. A.i Abyssal Plain

Abyssal Rock

Another name for plutonic rock or rock that has been formed by intense heat or pressure deep within the earth's crust.

Acapulco

A famous resort town in southwestern Mexico renowned for its scenic beauty and spectacular cliff diving feats by local inhabitants. These divers must time their fall to hit an incoming wave and insure enough depth at the base of the 120 foot high La Quebrada cliff face.

Acid rain

Precipitation polluted by acidic substances and falling as rain. Exhaust from cars and trucks, and pollutants that result from the burning of coal and other fossil fuels introduce deadly gases into our atmosphere. Sulphur dioxide and nitrogen oxides dissolve in rain drops to make acid rain and it damages forests and fishing grounds.

Acclimate

The process of becoming accustomed to a different climate, environment or altitude.

Acclivity

Refers to an upward incline; opposite to declivity.

Aconcagua

An Andean peak in western Argentina. At almost 23,000 feet above sea level, it is the highest peak in the western hemisphere.

Acorn

The fruit of the Oak tree; used to feed hogs.

Acre

A land measurement (43,560 square feet).

Adelaide

The capital city of South Australia, located on the southeastern coast. Basking in a Mediterranean climate, with warm dry summers and cool moist winters, this city is a manufacturing and marketing hub for the region.

Adirondack Mountains

This range of upland hills is part of the Appalachian chain that extends into northeastern New York state. It is a sporting and tourist center because of the abundance of wildlife and beautiful natural scenic vistas carved and sculpted by forces of glaciation.

Adriatic Sea

An arm of the Mediterranean between Italy and the former Yugoslavia. It is a popular tourist location hosting spectacular scenery, azure blue seas and a beautiful climate.

Fig. A.ii Adriatic Sea

Advancing - glacier

When the amount of snowfall high up on the slopes is greater than the amount of melt at the snout, the glacier is said to advance down into the valley.

Aegean Sea

An arm of the Mediterranean between Greece and Turkey. The islands of the Grecian Archipelago lie scattered in the Aegean Sea, which is connected to the Sea of Marmara by the Dardanelles and from there to the Black Sea.

Aeolian forces

Coming from Aeolus, the God of the winds, this word represents the forces of processes associated with the action of the wind. The resulting features, called ventifacts, are often quite spectacular and stunning. Wonderful examples of ventifacts can be seen in the Mid and Western states such as Utah, New Mexico and California, Bryce Canyon, for example. Other ventifacts are found in deserts and are known as yardangs and zeugens.

Aeronautical Charts

Topographic maps that are specifically designed to aid pilots and navigators in the business of flying their aircraft.

Aerosol

A small container in which pressurized gas is used to dispense a liquid through a valve in the form of a spray or a foam.

Fluorocarbons were used in the manufacture of the propellants for aerosol cans, but scientific research linked these chemicals with the diminishing ozone layer and their manufacture was banned by the U.S. Government.

Afghanistan

A country in southwest Asia between Iran and western Pakistan. Its capital is Kabul.

Africa

The second largest continent, situated in the eastern hemisphere south of Europe.

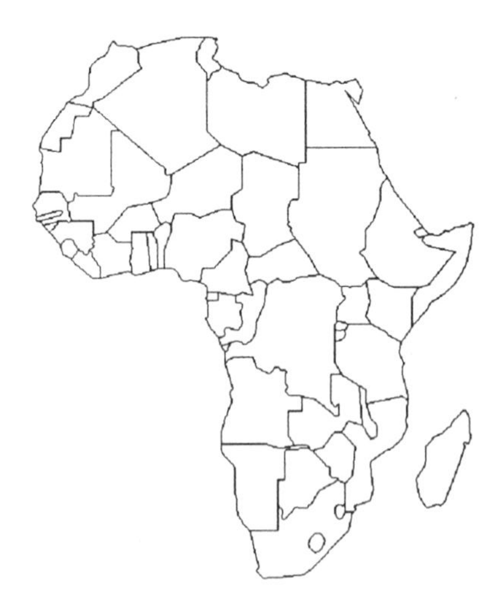

Fig. A.ii Africa

Afrikaans

One of the official languages of South Africa, developed from the 17th century Dutch people who settled this region. The other official language is English.

African

A native inhabitant of Africa, especially a dark skinned individual.

African rift valley

A remarkable valley that was formed by tectonic movement and stretches for over three-thousand miles, from Egypt in North Africa throughout East Africa and south to Madagascar. It is studded with dormant volcanic cones and spectacular lakes, including lakes Victoria, Nyasa and Tanganyika. It is also an area that is noted for its inhospitable living conditions in places, where the tsetse fly abound and temperatures and climatic conditions are decidedly difficult for sustaining life. See also; Rift Valley.

Afternoon

The time of day from noon to evening.

Aftershock

A minor earthquake following a greater

one and originating near or at the same place.

Agane

The capital city of the island of Guam.

Agate

A hard semi-precious stone with clouded or striped coloring as in playing marbles.

Agglomerate

A mass of fragments of volcanic rocks fused by heat.

Aggrade

A river valley or bed that is built-up by deposits of sediment and silt is said to be aggraded.

Agrarian

Relating to land, farmers, and agriculture.

Agriculture

The science and art of cultivating the land to grow crops and raise livestock.

Agronomy

The science and economics of crop production and management of farm land.

Aiguille

A term that is associated with the French Alps but is found in any alpine area. Aiguille is French for needle and refers to a peak of rock that is shaped like a sharp, sheer needle as a result of frost action.

Air

An invisible mixture of gases that makes up our atmosphere and that surrounds the earth. These gases include nitrogen and oxygen, as well as hydrogen, carbon dioxide, argon, neon, helium and others. Sometimes referred to as the sky, it is the space above the surface of the earth.

Air pocket

An atmospheric condition, sometimes called turbulence, that causes an air-plane to make sudden short drops while in flight.

Air pressure

The weight of the atmosphere bearing down on us.

Air masses

Bodies of air that are distinguished by differing temperatures. When these air masses meet, climatic conditions result. For instance, when a warm air mass meets and mixes with a colder air mass fog and precipitation may result.

Air pollution

Factories, automobile exhaust, effluent from our houses can contribute to the occurrence of pollution in our world.

Alabama

A southern state in the U.S. situated on the Gulf of Mexico. Its capital is Montgomery.

Alamo

A famous Franciscan mission at San Antonio, Texas. In 1836, this was the site of an historic siege and eventual massacre of a small band of Americans by Mexican troops.

Alaska

A U.S. state in the North American continent to the north of Canada and separated from Asia by the Bering Sea. It was bought from Russia, in 1867, for approximately 2 cents per acre. Alaska is rich in fish, minerals such as oil, timber, and other raw materials that have boosted the American economy.

Even though it is the largest state in the United States it has one of the smallest populations because of the ruggedness of the living conditions typically experienced there. Its capital is Juneau.

Fig. A.iii Alaska

Alaskan current

A cold current, originating in the Arctic waters off Alaska in the Bering sea, that moves south along the west coast of North America. It is responsible for frigid waters in lower latitudes and the occurrence of fog in places where the warm interior air comes into contact with the Arctic air over the current.

Albania

A country on the Adriatic Sea in the Balkan peninsula. Its capital is Tirana. See also; Adriatic.

Alberta

A prairie province in southwestern Canada that was named after the daughter of Queen Victoria. Its capital is Edmonton. Alberta is rich in natural resources and this accounts for the high standard of living enjoyed by the majority of Albertans. It has large oil fields, many coal mines and rich agricultural soils where beef and grain are produced in vast quantities. In addition, Alberta's natural scenic vistas attract thousands of tourists per annum to visit the national parks, including Banff and Lake Louise, nestled in the foothills of the Canadian Rockies.

Albuquerque

A city in New Mexico.

Alcan Highway

A famous stretch of highway that connects Alaska with the lower states across Canada.

Alcatraz

A small island in San Francisco Bay, now a tourist attraction, but once the site of a famous federal prison.

Alder

A tree or shrub of the birch family that grows in cool moist soil in temperate climates.

Aleutian Islands

A chain of islands extending 1200 miles southwest from the tip of the Alaska peninsula. See also; Alaska.

Alexandria

A seaport in Egypt, on the Mediterranean Sea, at the western end of the Nile delta. Founded by Alexander the Great, it was a seat of Hellenistic learning in the ancient world.

Alfalfa

A deep rooted plant of the legume family used extensively in the U.S. for fodder, pasture and as a cover crop to protect the soil.

Algae

A seaweed, pond scum or other plant commonly found in water or water-logged areas.

Algamest

A vast work (book) on astronomy and mathematics compiled by Ptolemy circa. 150 AD.

Algiers

A north African seaport on the Mediterranean Sea. It is the capital of Algeria.

Algonquin

A Native American tribe that inhabited the Ottawa region. Also refers to one of several languages that were spoken by some of the following tribes: Arapaho, Cheyenne, Blackfoot, Chippewa, Fox, Shawnee, Ottawa and others.

Alluvial fan

A gradually sloping mass of alluvium that widens out like a fan from the place where a stream slows down, little by little, as it enters a plain and its gradient changes. An alluvial fan occurs when a swift upland stream emerges abruptly into a level plain.

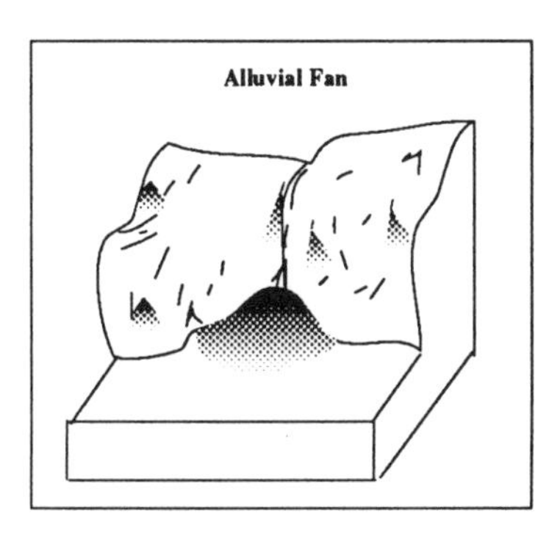

Fig. A.iv Alluvial Fan

Alluvium

Sand and clay gradually deposited by moving water on the bed, or floor, of a river or on the shores of a lake.

Almanac

A yearly calendar of days, weeks and months with astronomical data, weather forecasts and other statistical information.

Alp

A term that refers to the meadow area above the tree level but below the permanent snow-line that is used for grazing in summer.

Alps

A high mountain system in Europe that extends from southern France through Switzerland, Italy, Germany,

Austria, into the former Yugoslavia and Albania.

Alpine valley

A picturesque glaciated valley in an upland region.

Alternative sources of energy

Energy sources that differ from the traditional dependence on fossil fuels, like oil, coal and gas. Solar energy is an example of an alternative source of energy, as is wind or water-power.

Altiplanos

High-altitude plains in the Andes of South America.

Altitude

The height of a place above sea level.

Altocumulus

A formation of white or gray clouds in many shapes, at intermediate heights.

Altostratus

A formation of gray clouds at intermediate height in dense layers.

Amazon

A river in South America that flows over 3,300 miles from high in the Andes mountains in Peru across northern Brazil into the Atlantic ocean.

Fig. A.v Amazon Basin

Amazon Basin

An area in South America that the Amazon river system drains on its route down to the Atlantic ocean from its source in the Andes.

Amazon Rain Forest

The world's largest equatorial rain forest.

America

The land named after the Italian explorer Amerigo Vespucci. Today, its discovery is attributed to Christopher Columbus. The name was coined by Martin Waltseemuller, a German geographer, from Vespucci's Latin name - Americus.

American Falls

Part of the cascade that, together with

the Horseshoe falls on the Canadian side, make up Niagara Falls.

Amerind

The name given to an American Indian or Eskimo.

Amish

A strict religious rural sect based on the Mennonite faith. Today, most of the members of this religious sect pursue a rural, simple way of life in Pennsylvania and neighboring states.

Amphitheater

A level place surrounded by higher ground.

Amsterdam

Capital city of the Netherlands, situated on Ljsselmeer.

Fig. A.vi Amsterdam

Amundsen - Roald (1872 - 1928)

The Norwegian explorer who was the first man to reach the South Pole on December 14, 1911. He beat the British explorer Robert F. Scott to the pole by five weeks and accounted for his speed and success by his use of teams of dogs and sleds. Scott and his team, who used ponies, had to resort to pulling the sleds themselves as the ponies died from exhaustion. Having been defeated in their quest, Scott and his entire party perished of cold and hunger.

Amur

A very long river in Asia that forms the border between the USSR and China.

Anaconda

A South American snake of the boa family. It is a long, heavy reptile and can crush its victim in its coil.

Andes

The high mountain range that creates the backbone of South America from Peru to Tierra del Fuego. The highest peak is Aconcagua.

Andromeda Galaxy

A constellation containing the brightest and the nearest of the spiral nebulas.

Anemometer

A device for determining the speed of wind, commonly called a wind gauge. Sometimes an anemometer will also contain a gauge, or vane, for measuring the direction of the wind.

Aneroid Barometer

A partial vacuum box that is calibrated to measure minute changes in atmospheric pressure. Aneroid means "without liquid."

Angel Falls

A towering, spectacular waterfall on the Churun river in eastern Venezuela, South America, that drops in excess of 3,000 feet, into the valley below. It is the world's highest waterfall.

Angola

A country on the southwest coast of Africa. Most of Angola's people are rural dwellers and work at producing for home consumption or harvesting for export bananas, sugar cane, corn and coffee. Luanda is the capital city and is situated on a natural sea port.

Ankara

The capital city of Turkey.

Annapolis

A city on the Chesapeake bay; the capital of Maryland.

Annapurna

A Himalayan mountain block in Nepal.

Anno Domini

A.D. means literally - In the year of our Lord. It refers to how time is measured since the beginning of the Christian era.

Annular eclipse

When the sun is blocked by the moon but a ring is still visible around the rim. Annular means ring. An eclipse in which a ring of sunlight can be seen around the disc of the moon.

Anorak

Originating in Greenland, it is a heavy jacket with a hood worn in the cold northern climes.

Antarctica

The region of the globe that surrounds the South Pole and is characterized by ice sheets and barren tundra.

Antarctic current

The cold ocean current that originates in the Antarctic region and influences the temperature of the water and general climatic conditions in the southern seas.

Antarctic ocean

The body of water that surrounds the southern polar region and is rich in organisms and plankton that supports an abundance of marine life.

Antemeridian

A.M. means literally ante Meridiem - before noon, in Latin.

Anthracite

A hard, black layer of coal that is found in some sedimentary basins. It gives great heat with little flame and smoke.

Anthropology

The study of man, his distribution, customs and social relationships.

Anticlines

In fold mountains, an anticline is known as the up-fold and a syncline as the down-fold. An anticline is a sharply arched fold of stratified rock from whose central axis the strata slope downward in opposite directions.

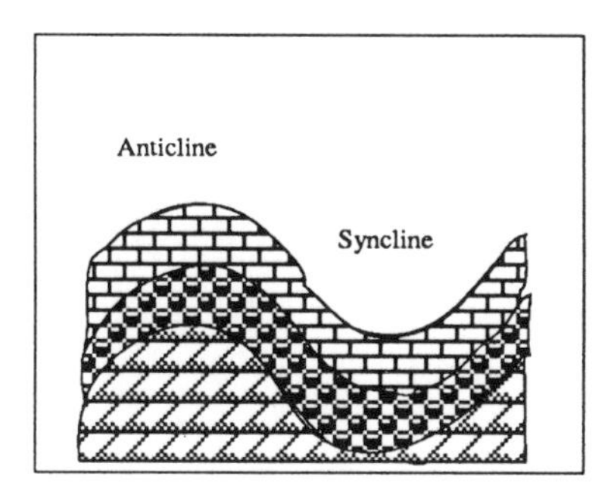

Fig. A.vi Anticlines

Anticlone

An atmospheric condition of high barometric pressure with the winds at the edge blowing outward.

Antwerp

A major port of Belgium, situated on the river Scheldt. Antwerp is renowned as a world class diamond cutting center.

Apennines

A mountain range running the length of Italy, forming the backbone.

Aphelion

When the earth is farthest from the sun in its orbital journey it is said to be at aphelion. It is the opposite to perihelion.

Apogee

The point farthest from the earth in the orbit of the moon. It is opposite to perigee. Since the moon's orbit is elliptical it is not always at the same distance from the earth, a fact that has a significant impact on the tidal movements.

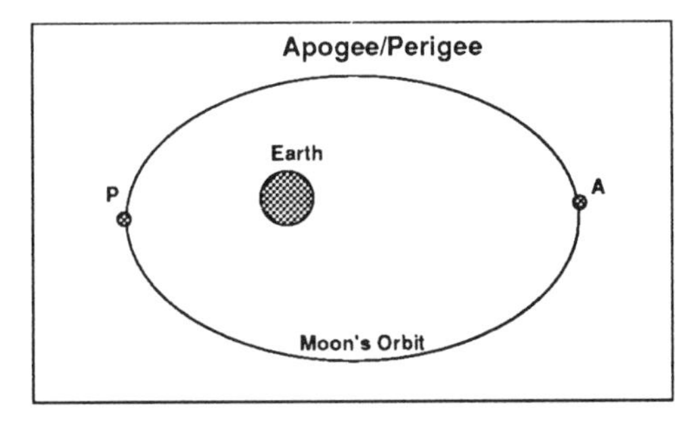

Fig. A.vii Apogee

Appalachian mountains

A range of high mountains in the eastern United States.

Aquaculture

The regulation and cultivation of ocean plants and animals for human consumption. It is also called saltwater farming.

Aqueduct

A method of transporting water by conduit, usually from distant sources to large cities or arid regions for human consumption or irrigation.

Aquifer

An underground layer of porous rock containing water into which wells can be sunk.

Arab

Inhabitant of Arabia, the desert region of the Middle East.

Aral Sea

An inland body of saltwater, in the USSR, east of the Caspian Sea.

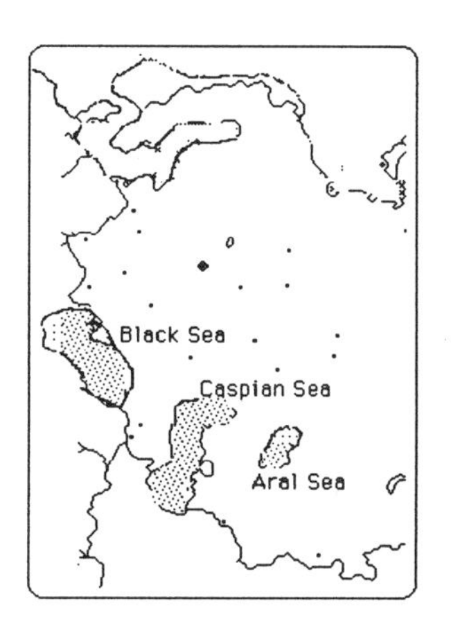

Fig. A.viii Aral Sea

Ararat

A 17,000 foot mountain in Turkey, near the border with Iran, that is reputed to be the landing place of Noah's ark.

Arboretum

A place where many kinds of trees and shrubs are grown for exhibition and/or study.

Archaeology

A research method for learning about past human cultures. Archaeologists study the objects, tools, bones, tombs and dwelling sites that ancient peoples left after them.

Archipelago

A chain of islands. The term was first used for the islands in the Aegean Sea.

Arctic Circle

An imaginary line that is 66.33° north of the equator and which defines the colder regions of the globe.

Arctic Ocean

The body of water that surrounds the North Polar regions of the globe. It is the smallest ocean in the world.

Arete

Knife-edged ridge that separates two pyramidal peaks in glaciated topography. Sharp narrow ridge or crest of a mountain. See also; Mountain ranges, Alps.

Argentina

A country in South America second in size and population to Brazil. Buenos Aires is the capital city. Most of the people live in cities. The country produces vast quantities of wheat and beef. Manufacturing industries are generated from the processing of these farm products.

Argon

A colorless, odorless gas that constitutes roughly one per cent of the atmosphere.

Arid

A dry desert region is said to be arid when there is not enough water for things to grow. Arid areas are not suited to agricultural activities and agronomy without the addition of water, a process called irrigation.

Arizona

A thriving state in the southwestern

U.S. Irrigation has transformed this arid region into an oasis of industry and tourism. Arizona is the state that is home to the Grand Canyon, the Painted Desert and the Petrified Forest. In the last few decades Arizona's population has grown enormously as more and more people are discovering the climatic and life-style advantages. Phoenix is the capital city.

Armenia

A mountainous republic between the Caspian and Black seas and bordered by Georgia and Azerbaijan. Yerevan is the capital city.

Armillary sphere

A model of the celestial sphere formed of fixed rings in the positions of the great circles, such as the tropics and the meridians.

Arno

A river in central Italy that flows west through Florence into the Ligurian sea. It is home to many species of gold fish.

Artesian well

A well that is drilled deep enough to reach through to water, from grounds higher up, that is flowing in bedrock so that the pressure will force it to rise to the surface.

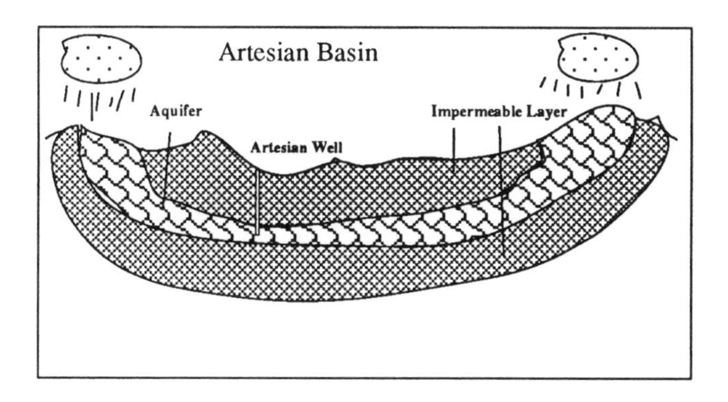

Fig. A.ix Artesian Basin

Ash (Volcanic)

Fine volcanic lava that results from an eruption.

Asia

The largest continent, bordered by Europe to the west, and situated between the Pacific and the Indian Oceans.

Asian plate

The tectonic plate between the "Pacific rim of fire" plate and the Himalayan plate.

Asp

A small European/African snake.

Aspect

The direction in which a slope is facing or its exposure, e.g. a house may have a southern aspect.

Astenosphere

A zone of plastic-like material below the harder lithosphere on the surface of the earth.

Astronomy

The science of the stars, planets and other heavenly bodies dealing with their motion, position, size and so on.

Asteroids

Numerous small planets that orbit between Mars and Jupiter.

Aswan dam

A large dam on the Nile that provides irrigation for the surrounding regions. It has greatly affected the ecological development of the area.

Astro-navigation

A form of navigation that uses the celestial bodies as aids to location.

Astronaut

A space traveler or explorer.

Astronomer

A person who studies astronomy.

Astrophysics

The science of astronomy as it applies to physics.

Atacama

A very large tract of desert in Chile, South America.

Athabasca, River

A river that rises in the Rocky mountain range in Alberta, Canada and flows into Lake Athabasca.

Athens

The capital of Greece on the Aegean Sea. This is one of the oldest and most historic cities in the world. Today it is a thriving center for tourism. Thousands of visitors flock to see the famous Acropolis (High City), the Agora (Market), and other ancient remnants of early Greek times.

Atlanta

The capital city of Georgia and a thriving center in the southeastern U.S.

Atlantic Ocean

The body of water between Europe and North America.

Atlas

A collection of maps and geographical information.

Atlas mountains

A range of mountains in North Africa extending across Morocco, Algeria and Tunisia.

Atmosphere

The invisible layer of air that surrounds our planet earth.

Atmospheric pressure

The weight of the air pressing down on the surface of the earth.

Atoll

An island made from coral that blooms in warm water especially in the south seas. A ring shaped coral island that surrounds a lagoon.

Atomic Energy

The energy released in nuclear fission or fusion.

Attrition

The wearing away of rocks and pebbles by friction from constant rubbing against each other, as occurs in the bed of a river or along a coast line.

Auckland

A seaport in North Island, New Zealand.

Audobon Society

A number of affiliated groups that promote the conservation of wildlife and natural resources.

Auk

Web-footed diving bird of the north seas.

Aurora borealis

Also known as the northern lights, these are streamers of lights that appear in the northern night sky, caused by electrical charges in the ionized air.

Australia

An island continent in the southern hemisphere, whose native inhabitants were Aborigines and that was colonized in the last century by the Europeans. The capital city is Canberra.

Fig. A.x Australia

Austin

The capital city of Texas named after Stephen Austin who formed a colony in Texas in the early 1800s. It is a market and manufacturing center for the surrounding agricultural hinterland, specializing in beef production. It also serves as a major educational and administrative center for Texas.

Austria

A European country to the east of Switzerland, famous for its spectacular Alpine scenic vistas. It has wonderful broad, green valleys between towering glacial peaks that are dotted with pristine finger lakes. Its capital city, Vienna, is a center for tourism because of its picturesque Danube setting and its many impressive historic buildings and rich culture.

Autumnal Equinox

The time in the northern hemisphere when the sun is opposite the tropic of Capricorn (September 22) and day and night are each equal in duration.

Avalanche

A sudden disturbance on a mountain slope that causes a slide of snow, ice or rock and can be catastrophic to people and to places in its path.

Axis

An imaginary central point about which the earth rotates once every 24 hours.

Azerbaijan

A Soviet republic on the Caspian Sea. Oil is the region's main source of wealth and many people work in oil-related industries. Caviar is processed from sturgeon caught in the Caspian Sea. Baku is the capital city.

Azores

A group of islands in the north Atlantic off the west coast of Portugal. They are part of the peaks of the underwater mountains of the mid Atlantic ridge.

Azov - Sea of

The northern arm of the Black Sea in southern European Russia.

Aztec

An advanced civilization that inhabited Mexico before the arrival of Cortes in the 16th century.

Azure

The color of the clear blue sky.

Backwash

Waves are forced headlong onto the shore by wind and tides and they retreat under their own weight. Swash runs up the beach and backwash falls back into the sea. The combined effect can cause longshore drift.

Badland

A region marked by powerful erosional sculpturing, scanty vegetation with steep hills, and deep gullies formed primarily by water erosion. Flash floods are the main cause of these features. Most badlands are not suitable for agricultural activities beyond grazing. Consequently the badlands are marked by local grazing and natural scenic habitats.

Baffin Bay

That part of the Atlantic that separates Baffin Island from Greenland, north of Hudson Bay, Canada. It was named after the English explorer of the seventeenth century who led an expedition to find the Northwest Passage to Asia. The island is the sixth largest in the world but is too remote to be more than sparsely populated.

Baghdad

Situated on the banks of the Tigris river, Baghdad is the capital city of Iraq.

Bahamas

A chain of about 3,000 coral islands and reefs situated in the West Indies between the tip of Florida and Cuba. The subtropical climate and scenic beauty helped establish the Bahamas as a tourist center for neighboring America. Its capital is Nassau.

Bahrain

This island country in the Persian gulf was catapulted to prominence during the Gulf War. Its importance lies in the fact that it has vast reserves of petroleum and its people enjoy one of the highest standards of living in the Gulf area.

Baja Peninsula

The narrow promontory that stretches below California into the south Pacific enclosing the Gulf of California. It is a Mexican semi desert region with barren mountains on the eastern side.

Fig. B.i Baja Peninsula

Bakersfield

A city in south central California.

Bald Eagle

A large and powerful bird of prey that feeds on fish and dead animals. Its white head feathers give it the appearance of baldness. Usually building their eyries on the tops of tall trees near water or on safe cliffs, they hunt during the day often soaring gracefully to great heights in the air. They have long been the symbol of freedom and power.

Balkan

A peninsula in southeastern Europe between the Adriatic and the Ionian seas to the west and the Aegean and Black seas to the east. The Balkan countries include Albania, Greece, Bulgaria, eastern Turkey and most of what was Yugoslavia.

Baltic Sea

A large inland sea in northern Europe separating the Scandinavian peninsula from the rest of Europe. Sweden's city ports of Copenhagen and Stockholm are situated on the Baltic sea. It is connected to the North Sea via the Kiel canal or through the Kattegat and Skagerrak waterways around peninsular Denmark.

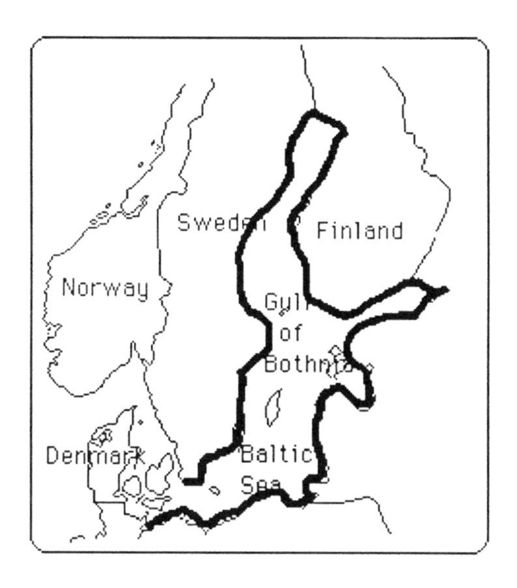

Fig. B.ii Baltic Sea

Banff

This city and national park in Alberta, Canada is noted for its spectacular natural beauty and is a major tourist center in the Rockies. It contains magnificent examples of glacial sculpturing and other erosional features making it one of Canada's singular attractions.

Bank, river

Sides of a river channel, sometimes undercut and likely to cave in when walked on.

Bangkok

The largest city in Thailand, a sprawling fast growing metropolis that has many canals as routeways (like Venice). It boasts many fine examples of oriental temples and palaces.

Bangladesh

An Asian nation, bounded on three sides by India. Bangladesh is one of the most overcrowded places on the globe, and is characterized by widespread poverty.

Barcelona

A seaport in northeastern Spain. It is a favorite European tourist center and was the site of the 1992 Olympic games.

Barley

This cereal crop like corn, oats and wheat is grown by farmers and is used for making malt and for feeding livestock. It thrives in temperate climates and can grow in all soils even at high altitudes. Where it is feasible, a winter crop is also sown if the climate is sufficiently mild to enable its growth to ripening.

Barn

A farm building used to house livestock and to store hay and other supplies used in agriculture.

Barograph

An instrument that keeps records of changes in the pressure of the atmosphere. It is used by weather scientists to forecast changes in the weather.

Barometer

An instrument for measuring atmospheric pressure. The mercury barometer responds to increases and decreases in atmospheric pressure by a corresponding rise or fall in the level of mercury in a graduated glass tube.

Barrier reef

A coral reef that extends the length of the eastern Australian coast and the watery resting place of numerous shipwrecks. It is also known as the Great Barrier Reef.

Basalt

A hard, dark extrusive volcanic rock that forms when lava flows from its eruptive origin - either a lava cone or a rock fissure - and cools. Sometimes it forms spectacular formations, e.g., giant cliffs of octagonal columns, or it may solidify to create gigantic plateaus. It is the most common of the volcanic rocks. Hawaii is mostly formed of basalt.

Basel

A city in Switzerland, on the river Rhine bordering Germany.

Base Level

The level below which a stream cannot erode its bed.

Base Line

A horizontal line measured with extreme accuracy to assist in surveying for triangulation.

Base Map

An outline map, upon which data may be plotted.

Basque

A region and people inhabiting the Pyrenees on the border between France and Spain. These culturally unique people have enjoyed limited self rule since 1980.

Bass Strait

The body of water that separates Australia from Tasmania, 80 to 150 miles wide.

Batholith

A great volume of intrusive igneous rock that forms when magma hardens within the crust. The magma stops on its course towards the surface and solidifies at a considerable depth below the ground. Batholiths usually form the core of mountain ranges, and since the magma cools slower beneath the surface than on the outside, large crystalline rock formations, for example granite, result.

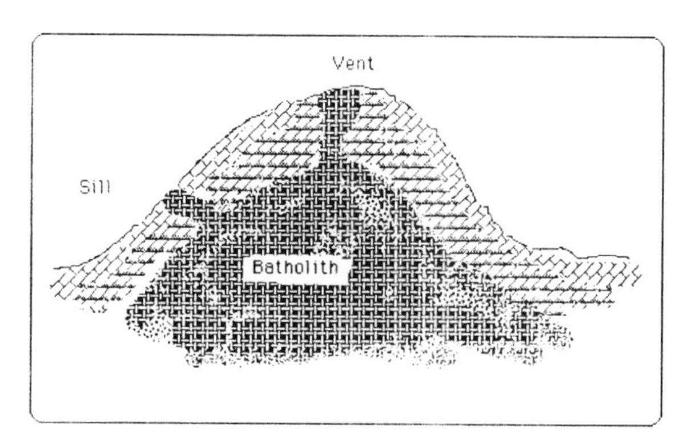

Fig B.iii Batholith

Baton Rouge

The capital city of Louisiana and a major port on the Mississippi.

Bay

An indentation in the shoreline that is surrounded by headlands or capes. See also; Headland.

Bay of Fundy

This arm of the Atlantic Ocean between New Brunswick and Nova Scotia is famous for the world's highest tides. The peculiar configuration of the shoreline and the direction of the North Atlantic Drift ocean current allow tides that are often in excess of 50 feet. Canadian engineers are considering harnessing the mighty power of the tides to generate electricity.

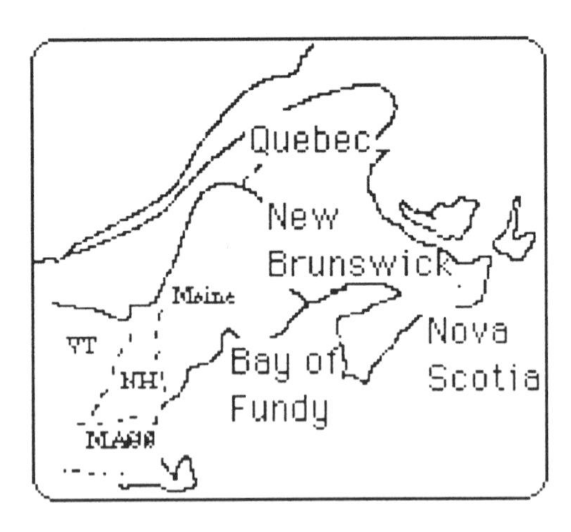

Fig. B iv Bay of Fundy

Beach

The area of shore between a cliff and the water mark. It can be a sandy, muddy, or pebbly edge to a water body.

Beach cross-profiles

A theodolitic survey of a beach to map the gradient and salient features between the cliff and the water-line. This method of learning is useful for class instruction in field activities. A typical beach profile could contain a cliff face, a lagoon, a storm beach, beach cusps, shingle, sand, ripple marks, water pools, runnels and ridges.

Bearing

The direction or situation of a point or place with regard to another or to a compass.

Bed, river

The channel that contains the body of water known as a river.

Bedouin

A nomadic desert tribe of Arabia, Syria or North Africa.

Bedrock

The solid rock beneath the soil.

Beech

A large forest tree that is found in North America and in Europe. Its leaves turn a wonderful golden hue, that augments the natural beauty of any landscape, in autumn.

Beef

One of the chief foods in many countries, it is the meat obtained from mature cattle.

Beet

A root crop that can be used in the manufacture of sugar or as a fodder crop for livestock.

Beijing

The third largest city in the world is the central city of Chinese culture and the capital city of China. It is also known as Peking, and dates back more than 2,000 years as the seat of government of several Chinese dynasties. There are many beautiful palaces and temples in Beijing as well as art treasures and universities.

Beirut

The capital city of Lebanon and a sea port on the Mediterranean.

Belfast

The capital city of Northern Ireland. It is an industrial and marketing center.

Belgium

A small country in northwestern Europe, whose geographic location has brought it much involvement in European politics and history in the past and makes it an important central point in the European Community today. One of the most densely populated countries in the world, it is a European center of international economic and political activity.

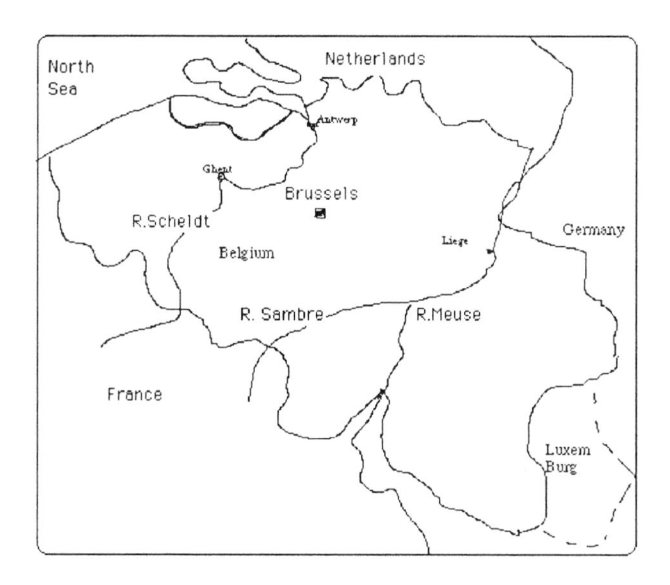

Fig. B V Belgium

Belgrade

The capital city of Serbia on the confluence of the Danube and Sava rivers. Its strategic location has turned it into a river port and railroad center.

Belize

A small developing country in Central America. It is located on the southeastern side of the Yukatan peninsula. The capital city is Belmopan.

Bend, river

Rivers turn and meander because of internal friction resulting in river bends.

Bengal, Bay of

A large bay in the northern part of the Indian Ocean bordered by India and Burma. Three major rivers spill into the Bay; the Ganges, the Brahmaputra and the Irrawaddy.

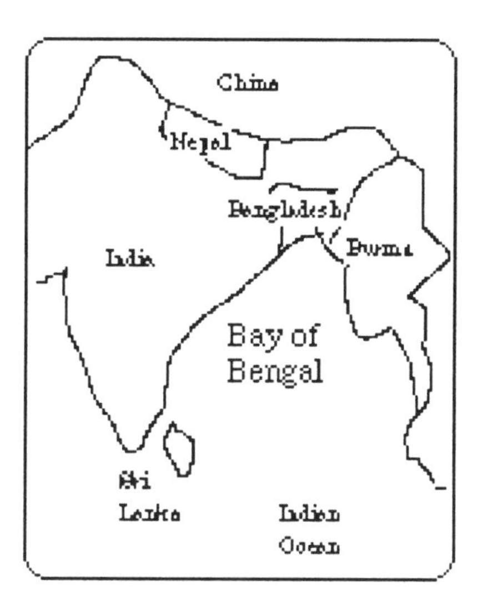

Fig. B V Bay of Bengal

Bering Sea

The body of water that separates Alaska from Russia.

Berkeley

A university city in California, on San Francisco Bay, called after the Irish philosopher.

Berlin

A city in Germany. It was united in 1991 after a long period of separation brought about by a militarily enforced wall. It was erected in 1961 to halt communist, East Germans emigrating to the West in search of better standards of living.

Bermuda

A cluster of approximately 300 small coral islands in the Atlantic, east of North America.

Bermuda Triangle

The body of water near the Sargasso sea where many strange events are reputed to have taken place.

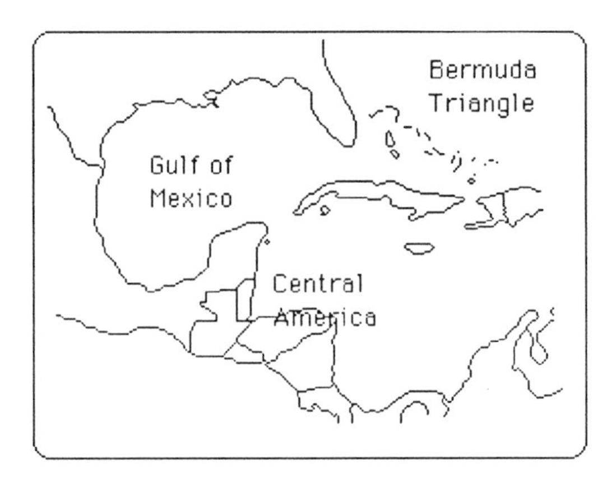

Fig.B.iv Bermuda Triangle

Bern

The capital city of Switzerland on the river Aar.

Big Bang theory

The theory that tries to explain the origin of the planet earth as a result of a cosmic explosion that hurtled celestial bodies into different orbits in space/ time.

Big Dipper

A constellation in the northern sky. See also; North Star.

Biodegradable

Capable of being readily decomposed by biological means.

Biogeography

That branch of geography that deals with the geographic distribution of plants and animals.

Biomass

Plants and wood can store solar energy which we can convert into fuel as sources of energy. For instance, plants that contain starch can be converted into alcohol which in turn can be burned to run tractors and other engines for industrial purposes.

Biosphere

A zone containing all the living organisms of the earth.

Bird's foot delta

The peculiar depositional formation that occurs at the confluence of river and sea, when the river has to stop and drop its load. The streamlets or distributaries that weave through the mud-flats often resemble the impression of a bird's foot, when viewed from overhead.

Birth rate

The number of births per year per thousand of population in a specific community.

Biscay

A bay in the Atlantic Ocean north of Spain and west of France. It is renowned for being a stormy place and proves a hazardous crossing for shipping.

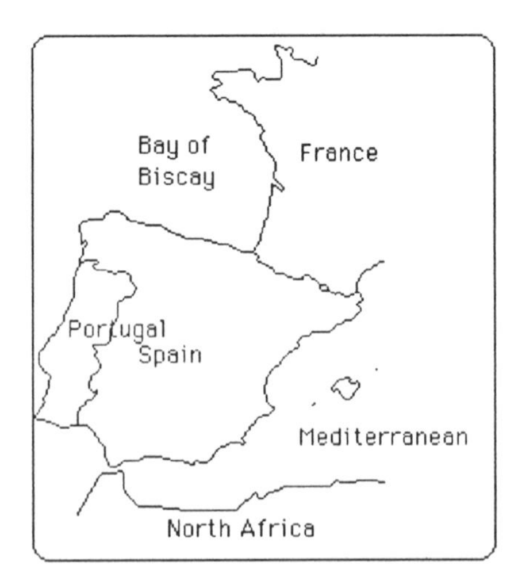

Bituminous coal

A black, soft coal that burns with a lot of dirt and smut, but does not give off as much heat as anthracite.

Black Forest

A wooded mountain region in south-western Germany.

Black Hills

The name given to a mountainous and wooded, badland region in Dakota, in the western U.S.

Black holes

The phenomena in space, that are determined to be, but cannot be seen. Their presence is deciphered by the reaction of other bodies to them but we cannot directly examine them. An area of such intense gravity that even light cannot escape and therefore cannot be seen by the naked eye.

Black Sea

A large water body near the Balkan peninsula in European Russia. It is a strategic connection to the Mediterranean for inland Soviets, via the

Bosporus Strait, the Sea of Marmara and the Dardenelles. Many large rivers empty into the Black Sea, including the Danube, the Dnieper, the Dniester and the Don.

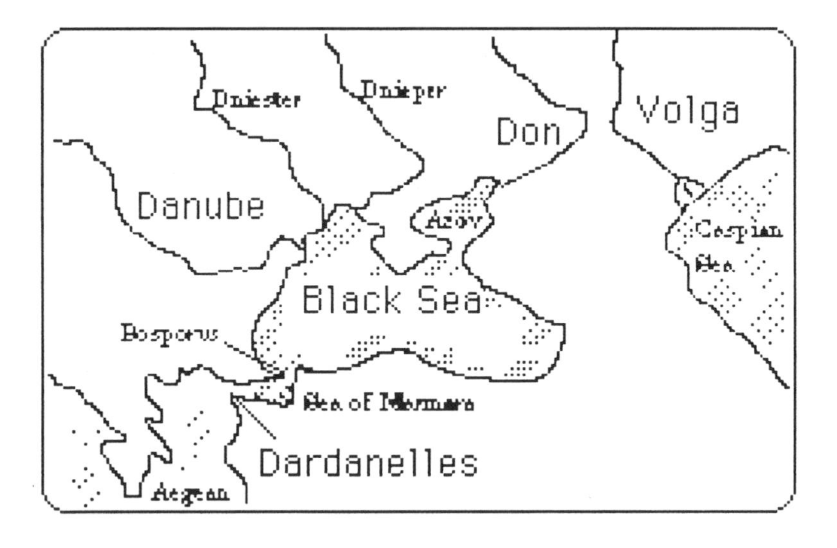

Fig. B.v Black Sea

Blanc, Mont

The highest peak in the European Alps at 15,781 feet. It is situated on the French, Swiss and Italian borders.

Mont Blanc

Blarney Stone

A stone atop Blarney Castle, County Cork, Ireland, which is said to impart the "gift of eloquence" to anyone who kisses it.

Blizzard

A violent storm with powdery, driving snow and extremely cold winds.

Block Diagram

A three dimensional perspective representation of geologic or topographic features showing a surface area and usually two vertical cross sections.

Block Mountain

A mountain produced by faulting and uplifting of large blocks of rock.

Blow Hole

A crack in a rock along a shoreline, through which wave action will sometimes force water upwards in a spectacular spout.

Blue Ridge Mountains

The easternmost range of the Appalachians extending from southern Pennsylvania to northern Georgia.

Bluff

A promontory of land ascending steeply with a broad flat front.

Bog

A small marsh or swamp with soft moist ground. Peat is cut from a bog.

Bohemia

A region in eastern Europe. In modern parlance a Bohemian life-style is resembled to that of a gypsy, and suited to poets, writers and artists.

Boise

The capital city of Idaho in western U.S.

Bolivia

A land-locked country near the center of South America. Most Bolivians are rural dwellers and subsist at a low standard of living because of the harsh climate and inhospitable terrain.

Bombay

The largest city of India, situated on an island on the western coast, is a thriving sea-port exporting cotton and other goods. Overcrowding in the cities and conditions of dire squalor are a reality in modern India for the vast majority of its inhabitants.

Bordeaux

A seaport in southwest France on the river Garonne. Also an agricultural region in southern France with famous wine producing vineyards.

Boreal

A northern zone of plant and animal life lying below the Tundra and characterized by coniferous forests and northerly winds.

Borneo

A large island in the Malay Archipelago southwest of the Philippines. It is the third largest island in the world.

Bosnia and Hercegovina

A country in southeastern Europe, formerly part of Yugoslavia. The capital city is Sarajevo.

Bosporus

A strait in southeastern Europe connecting the Sea of Marmara and the Black Sea. It is an important shipping route, providing a link to the Mediterranean Sea. See also; Black Sea.

Boston

The capital city and a university town in Massachusetts.

Botany

The study of plant life.

Botany Bay

A bay on the southeast coast of Australia, near Sydney.

Botany Bay, Australia

Bothnia, Gulf of

An arm of the Baltic Sea between Finland and Sweden. See the map of the Baltic Sea.

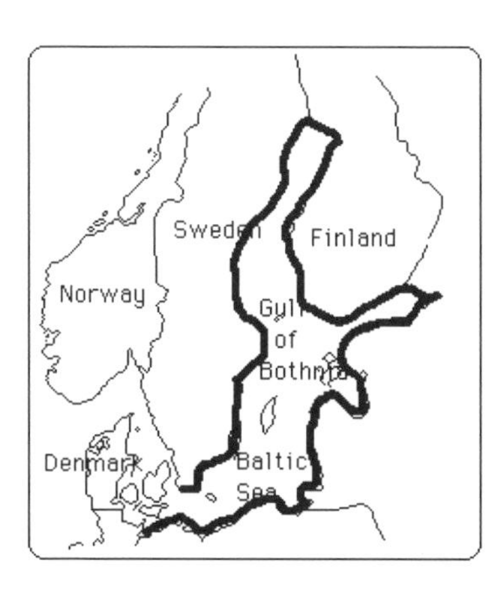

Fig. B.v Gulf of Bothnia

Boulder clay

The glacial till that results in the wake of an ice age when depositional material is dumped on the landscape. Rich soils sometimes develop in these areas.

Braided stream

When a stream in its upper course is divided and flows in numerous interconnected channels it is said to be braided.

Brazil

A country in northeastern South America on the Atlantic. The river Amazon flows through Brazil. Brazil is the location of vast rainforests which are being depleted today. The capital city is Brazilia. See also; Amazon Basin.

Breakers

Storm waves that crash onto the beach are known as breakers since the circle of the wave is broken and the water is destructively tumbled onto the surface of the beach. Breakers are distinguishable by their foamy line of surf on top of the rollers as they approach the shore.

Break water

A barrier, usually man-made, to lessen the impact of waves on the landscape.

Brie cheese

A ripened soft white cheese named after the region east of Paris, France, where it is made.

Britain

The island country in the northwestern corner of Europe that encompasses England, Wales and Scotland as well as six counties of N. Ireland. Also known as Great Britain.

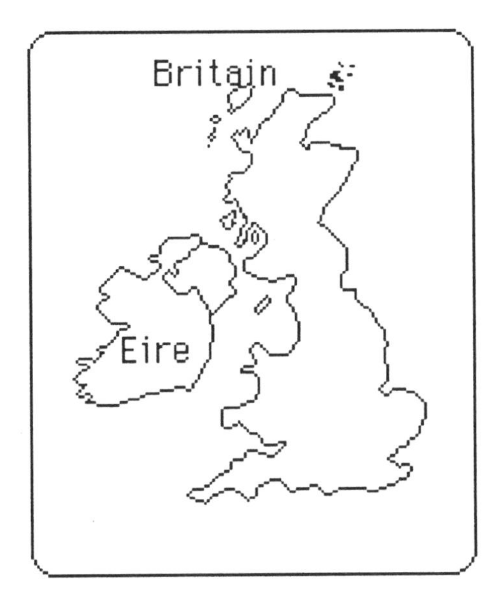

Fig. Bvi. Britain

British Columbia

The western-most Canadian state, bordering the United States. The capital city, Vancouver, is the chief Pacific port of Canada and the cultural and economic center of the western region.

Brittany

A region of northwestern France between the Bay of Biscay and the English Channel. It is more Celtic than French and has close cultural links with Wales and Ireland. Most of the people are engaged in fishing or tourism in this idyllic but remote corner of Europe.

Brontosaurus

A huge dinosaur of the Jurassic era that was up to 75 feet long.

Brow

The name given to the projecting top edge of a steep hill or cliff.

Brunei

A small country on the north shore of the island of Borneo in Southeast Asia. It is a country of considerable wealth because of the abundance of offshore oil reserves. The capital city is Bandar Seri Begawan.

Brussels

The capital city of Belgium and the seat of government for the European Community.

Bryce Canyon

A spectacular geologic phenomenon and national park in southwestern Utah.

Budapest

The capital city of Hungary situated on the banks of the Danube. It is the most important city in Hungary and the center of banking and government.

Buenos Aires

The capital of Argentina in South America.

Buffalo

A city in western New York State on Lake Erie, near Niagara Falls. Also a wild oxen or American bison.

Bulgaria

A country bordering the Black Sea in the Balkan peninsula. It is a mountainous land and its people struggle to find employment under adverse political and economic conditions. The capital city is Sofia.

Burma

A country bordering the Bay of Bengal in South East Asia. The river Irrawaddy empties into the Bay of Bengal through the many distributaries of a huge delta where Rangoon, the capital city is built. The Burmese people eke out a scant existence by farming the land.

Burton, Sir Richard F. (1821-90)

An English explorer of the Nile region of Africa and other locations.

Bushman

A person who lives in the Australian outback or a member of the nomadic tribes that occupy the Kalahari desert of southwestern Africa.

Butte

A steep hill standing alone on the plain. A small mesa.

Cactus

A desert flower, usually with fleshy skin and with spines instead of leaves. This plant can survive in arid regions since it stores water in its skin for long periods of time.

Cairo

The capital city of Egypt, situated on the Nile delta in northern Africa.

Cajuns

Descendants of French speaking settlers who were forced to flee Acadia and Nova Scotia and eventually settled in the Bayou area of Louisiana. They foster, to this day, a unique culture and tradition that is a blend of their French,

American and African American backgrounds. Their foods are spicy and include a soup called gumbo and a rice dish called jambalaya. They play a lively dance music that symbolizes their cultural heritage.

Calais

A seaport on the French side of the English channel, opposite Dover. The new channel tunnel, sometimes called the Chunnel, will focus more traffic in Calais and insure more shipping trade between England and mainland Europe.

Calculus

The science of mathematics first discovered by Newton.

Calcutta

The capital city and chief port of West Bengal in eastern India. It was founded in 1690 by the East India Company, an English trading firm, and at one time was a top ranking city in the British Empire. Today, it is one of the largest city in India and the majority of its three million inhabitants live in squalid conditions in shanty towns or sleep in the streets. Many work for small wages in the jute factories and most cannot read or write.

Caldera

A volcanic crater that sinks back into its cone forming a spectacular cone within a cone feature.

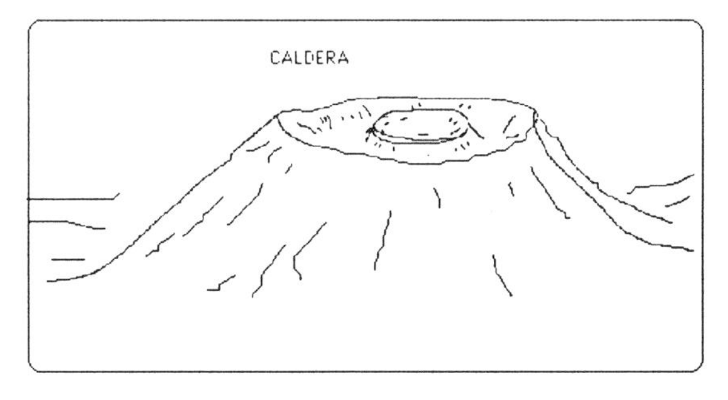

Fig. C.i Caldera

Calendar

A system for arranging the days and weeks of the year into an ordered and pragmatic format.

Calgary

A large and prosperous city in the eastern foothills of the Canadian Rockies, in Alberta. It grew rapidly from its beginnings as a cattle town to a booming petro-chemical center and today it prospers from its strategic location in the heart of agricultural

Alberta and is surrounded by many natural resources. In the 1970s the downtown area was refurbished and it now has the look and feel of a thriving modern city.

California

A state on the west coast of U.S. Sacramento is the capital city. California is one of the most popular states since it attracts visitors and inhabitants with its favorable climate, outdoor way of life and comfortable living conditions. The state is rich in raw materials, soils and minerals. As a result, California grows most of the fruit, and vegetables for the United States. These favorable conditions also attracted industrialists and employment in aircraft manufacturing, film making and computer industries has been considerable. Millions of people live in the coastal cities, San Francisco, San Diego and Los Angeles.

Californian Condor

Largest bird of prey in the U.S. An endangered species today.

Cambodia

A country in southeast Asia bordering on the Gulf of Siam. The capital city is Pnomh Penh.

Cambridge

One of the two oldest university cities in the U.K. It gets its name from a crossing point on the river Cam north of London. It is noted for learning and scholarship in nuclear physics and modern literature.

Camel

A large animal that can travel great distances over arid lands, fortifying itself for the journey with stores of food and water in its hump.

Cameroon

A republic in west central Africa. The capital city is Yaounde.

Camp David

A secluded retreat in the Maryland uplands, for the president of the United States. It was called after President Eisenhower's grandson David, in 1953.

Canada

The territory on the northern border with the U.S. of British colonial origin. Ottawa is the capital city. It is the second largest country in the world, but is thinly populated because of the rugged terrain and harsh climate that extends over vast areas of the northern lands. It has large stocks of natural resources by way of minerals, rich soils and scenic landscape and its inhabitants enjoy a high standard of living.

Canal

A man-made waterway that is useful for navigation and water conveyance (aqueducts). Water can be transported in canals to introduce irrigation to arid regions or to drain wet regions. The construction of dams and locks enable vessels to travel over portions of rivers that would normally pose navigation hazards like shallows, waterfalls and currents.

Canary current

A cold body of water that gets its name from the Canary islands that it passes by on its way back to the equator.

Cannes

A trendy resort town in the French Riviera, and site of a major international film festival.

Canberra

The capital city of Australia situated on the Murrumbigee river southwest of Sydney.

Canyon

A steep sided gorge resulting from erosion by agents of physical and chemical weathering.

Cape

A prominent headland surrounded on three sides by water. This coastal feature is formed either by erosion or deposition or by a combination of both.

Cape Canaveral

Situated on the east coast of Florida, it is the center of NASA's space operations where most of the missions, rockets and shuttles are launched.

Cape Horn

The most southerly point of South

America, is named after the town, Hoorn, of the Dutch sailor/explorer Willem Schouten.

Cape of Good Hope

A peninsula near the southern tip of Africa. It was named after Diaz voyaged as far south as this, in the hope that it would prove to be a routeway to India.

Capes and Bays

The long lists of geographical names that school-going children were forced to memorize long ago.

Capture, river

River capture occurs when the topography allows it. The tributaries erode faster than the main streams and cutting down and backwards at a faster rate, they capture the main channel. Often, features called wind gaps (dry valleys without rivers) are abandoned high on the hillsides where former rivers used to flow.

Carbon-14 dating

A system that scientists use to date rocks and archaeological finds from the past.

Carbon dioxide

A gas that is emitted when humans and animals breathe out. Green plants take carbon dioxide from the air and give off oxygen when light shines on them.

Carbon monoxide

The emission gas of automobiles.

Carrara marble

In Italy, near a town called Carrara, this fine marble is quarried.

Caribbean Sea

The sea near the Gulf of Mexico dotted with beautiful islands and lagoons. It is situated between the West Indies and Central and South America.

Carson City

The capital city of Nevada on the shores of Lake Tahoe.

Cartography

The science of map-making.

Cascade Range

The mountain range between the Rockies and the Pacific coast of the northwestern United States. It is characterized by plate tectonic movement and the amalgamation of two plates; the North American plate and the Pacific plate meet in a subduction zone that causes vulcanicity and earth tremors. There are numerous large volcanic cones along the ridge of this range, stretching from the Canadian border into California.

Caspian

An inland sea between Europe and Asia. It is a salt lake about 85 feet below sea level and the largest inland body of water in the world. Many rivers empty into the Caspian, including the Volga and the Ural. But in recent centuries its size is shrinking since evaporation from the lake exceeds input. Irrigation in the Volga drainage basin has used up much of the river's potential and the effects are felt in the Caspian.

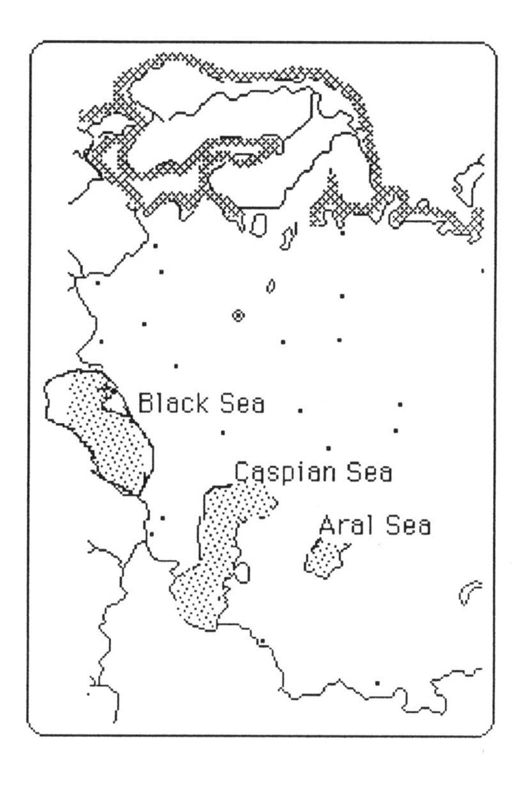

Fig. C.ii Caspian Sea

Catchment area

The region of territory that is drained by a particular river and its tributaries.

Cave of the Winds

At Niagara Falls there is a hollowed-out feature known as the *"Cave of the Winds"*. It was formed by differential erosion of hydraulic action in the beds of shale, slate and limestone.

Caves

Also known as caverns, these features can be found inland or on the coast. Inland caves are usually formed by chemical weathering in limestone areas. The features usually found on shorelines are typically the result of hydraulic and chemical erosion by waves.

Celsius

A metric scale for measuring temperature. On this scale zero is the freezing point for water and one hundred degrees is its boiling point. Since the scale is then divided up between these fixed points into a hundred parts, it is also known as the Centigrade scale.

Celts

A people and language that descended from the Indo-European culture. Today, Celtic areas include the western seaboards of Europe, typically, Galicia, Brittany, Wales, Scotland and Ireland. These regions share common traits in their language, culture and music. There is a large body of myths and legends associated with the Celtic race.

Cenozoic era

In geologic time the period before the Mesozoic era. This was the age when mammals were believed to develop, roughly sixty-five million years ago.

Census

A gathering of information by the government about the people and conditions throughout the land.

Central America

The narrow land connection between North and South America that contains Guatemala, El Salvador, Honduras, Nicaragua, Costa Rica, Belize and Panama.

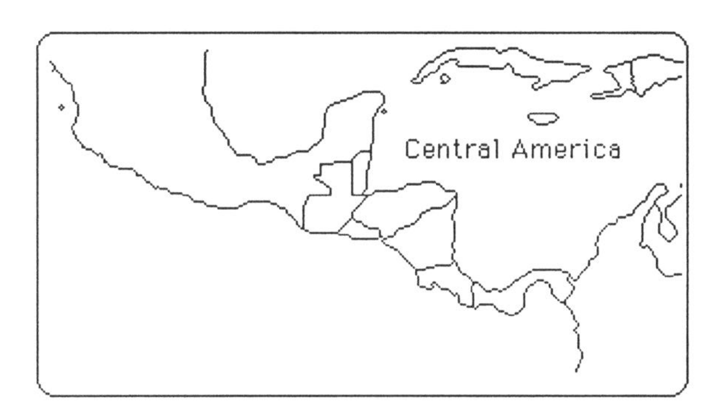

Fig. C.iii Central America

Ceylon

The former name of an island in the Indian Ocean, off the southern coast of

India. Colombo is the capital city. It is now called Sri Lanka.

Chaco

A semi-arid plain in South America shared by Paraguay and Argentina. It is populated mainly by cattle ranchers and is drained by the river Paraguay and its tributaries.

Chad

A desert republic in north central Africa. The capital city is N'Djamena.

Chalk

A sedimentary rock of fine white substance. It is found extensively in southern England where it has made famous the white cliffs of Dover.

Channel

The course of a river.

Chemical weathering

The erosional ability of plants and the atmosphere to wear away rock by the application of chemical elements and solutions.

Chernobyl

The site in the former Soviet Union where a nuclear disaster occurred in the 1980's.

Chesapeake Bay

A navigable channel in the eastern coast of the U.S. Many large rivers empty into Chesapeake Bay including the Potomac and the Susquehanna. As a result, there are important ports in Virginia (Baltimore) and in Maryland (Norfolk).

Cheyenne

The capital city of Wyoming.

Cheyenne Indians

Native American people now living mostly in Oklahoma or Montana. The Cheyenne took part in the Battle of the Little Bighorn, in Montana, when General Custer was defeated. Today, many Cheyenne have moved to cities.

Chicago

A thriving urban population inhabits this city, the third largest in the U.S. situated along the picturesque southwestern shore of Lake Michigan. Because of its strategic location, on a navigable waterway and adjacent to the vast agricultural lands to the west, Chicago became a leading industrial and transportation hub of the United States.

Chile

A long narrow country of South America between the Andes mountains and the Pacific Ocean. Santiago is its capital city.

Fig. C.iv Chile

Chimney

A narrow steep rock crevice that is climbed by adventuresome rock climbers.

China

The largest country in east Asia, also known as the second world. Peking or Beijing is the capital city. Roughly one-fifth of the world's population lives in China, but most of them live in the eastern one-third, due to rugged terrain and inhospitable climates. Its vast domain includes some of the driest deserts and highest mountains of the world.

Cinder cone

A direct result of volcanic activity, cinder cones build up over progressive emissions of lava and ash.

Circumference of the earth

The mileage around the globe first measured in 100 AD by Eratosthenes and found to be roughly 8,000 miles.

Cirrus

A high wispy cloud characterized by ice crystals and usually heralding the advent of a warm front.

City

A place where communities of people congregate to find work and set up their homes. Cities grow up on locations for many reasons. All cities need a plentiful supply of water and proximity to a supply of food. Advantageous places include crossing points on rivers, safe deep harbors and natural trading places. Modern cities continue to attract more and more people in spite of seemingly adverse conditions like overcrowding, violence and pollution.

Clay

A mixture of sand and silt.

Cliff

Steep rock formation usually found near a shoreline as a direct result of the erosive action of the sea.

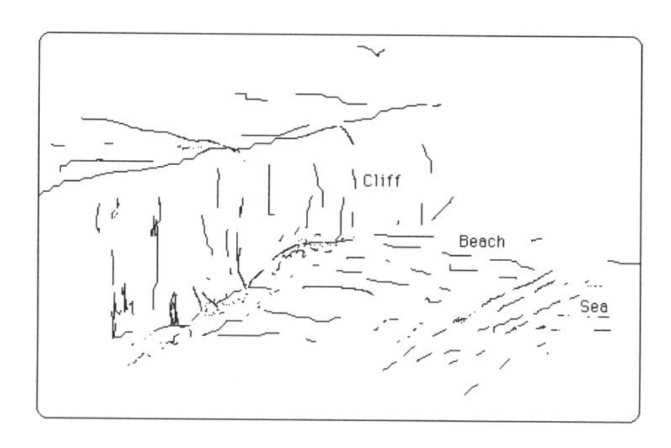

Fig. C.iv Cliff

Cliff under-cutting

The full power of wave action is only effective low on the cliff face so that undercutting occurs. The cliff is prone to collapse over time, as the base is hewed away systematically storm after storm.

Fig. C.iv Cliff

Climate

The accumulative weather pattern averaged over a period of about thirty years.

Cloud type

- Cirrus

A high wispy cloud characterized by ice crystals and usually heralding the advent of a warm front.

- Cumulo-nimbus

A low dark accumulation characterized by rising warm air and usually heralding the advent of rain. The Latin word for rain is nimbus.

- Nimbo-stratus

A high stratified cloud characterized by dark accumulations and usually heralding the advent of cloudy wet weather.

- Strato-cumulus

A high stratified accumulation characterized by rain clouds and usually heralding the advent of bad weather.

Coal

A dark, hard sedimentary rock that can be mined and used as fuel to generate heat and power.

Coast

The place where the sea meets the land, also called seashore.

Cold front

When warm and cold air masses meet frontal patterns develop. A cold front brings bad weather, with showers and snow. These fronts usually originate in the colder latitudes and move in to replace the warmer, rising air near the tropics.

Colombia

A country located on the northwestern corner of South America. Colombia has a variety of climates because of its situation on the equator and because much of its terrain is spread over high Andean peaks. It also has a coastline on both the Atlantic and Pacific oceans, separated by Central America. Bogota is the capital city of Colombia. Many inhabitants make their living growing coffee and from manufacturing industries spawned by the agricultural hinterlands.

Colorado, river

A river in the southwestern U.S. that rises in Colorado state and flows into the Gulf of California. The Grand Canyon, on the course of the Colorado, is one of the most spectacular effects of erosion and isostasy and thousands of visitors come to experience this delightful vista each year.

Colorado, State

A state in the southwestern U.S. whose capital city is Denver. Colorado is a tourist haven with some of the most picturesque landscape in the United States in Aspen, Estes Park and Colorado Springs. Mining and agriculture are also major sources of income for Colorado's growing population.

Columbia River

A river that rises in the Canadian Rockies, drains many states in the Pacific Northwest and spills into the Pacific Ocean at Astoria, on the border of Washington and Oregon. The Columbia River is the primary source of hydro-electricity, in the contiguous U.S. for the Pacific Northwest.

Columbia River Gorge

The deep and picturesque ravine that resulted when a glacial spillway formed a wide, deep channel during the last ice age. The Columbia River now forms the border between Oregon and Washington for much of its course.

Columbus

The capital city of Ohio.

Columbus, Christopher

Reputed to have first discovered America, he set sail from Portugal in the year 1492 on board the Santa Maria and made out the West Indies.

Comet

A celestial body that follows an elliptical orbit that places it close to the sun and our planet every so many years. Halley's Comet, probably the best known comet, orbits the sun roughly every seventy seven years. The last orbit occurred in 1986. A comet's tail always points away from the sun, a result of the direction and force of solar winds.

Commonwealth of Independent States

Formerly the Soviet Union, this is a collection of independent nations that are linked together geographically and politically.

Commonwealth of Nations

Countries that have, in the past, had colonial ties with Britain are linked by their common tradition and history into a loose association for economic and social cooperation.

Compass

The directional instrument calibrated in degrees, minutes and seconds. North always points to magnetic north.

Composite volcano

A volcano that builds up over progressive eruptive cycles, with cinder cones and lava flows.

Condensation

When warm air rises it cools and condenses forming fog and clouds.

Conglomerate

The amalgamation of rocks, pebbles and sand that form new rock, over time, by becoming cemented together in a chemical reaction.

Conservation

A practice of managing the Earth's natural resources wisely to insure their lasting supply. Conservationists strive to educate governments and people about the dangers of over exploiting nature and destroying the environment. Renewable and non-renewable resources are the focus of this thrust. Water, air, timber and soil are among the resources that are deemed precious to mankind and worthy of conservation.

Continent

A part of the Earth's crust that forms a land mass high and dry above sea level, but mostly surrounded by oceans. The continents are Europe, Asia, Africa,

North America, South America, Antarctica and Australia.

Continental drift

The belief that the continents were at one time one globular unit and that they drifted apart over time - and they are still drifting.

Continental ice sheet

The large bodies of ice that sit on the polar regions of our planet.

Continental shelf

The shallow shelves that project underwater from the coastline of most continents. Sunlight penetrates these shallow waters to produce rich harvests of plankton for the regeneration and maintenance of marine life.

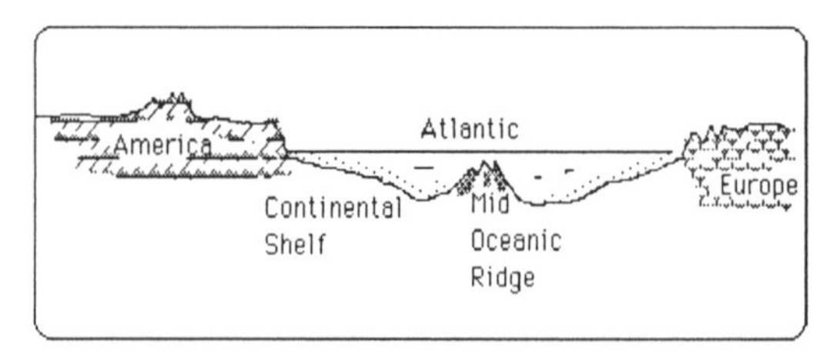

Fig. C.v Continental shelf

Contour lines

Imaginary lines that are drawn between places of equal height on a map. These lines enable the user, with a little training, to tell the gradient and read the terrain.

Contour ploughing

A method of ploughing devised to halt soil erosion resulting from runoff and soil creep. The farmer ploughs around the hill as opposed to up and down the hill.

Convection current

Air movement where hotter air rises and colder air flows in to replace it.

Cook, James

An eighteenth century explorer who was the first European to sail to Hawaii and Australia.

Coral reef

A limestone growth in warm water conditions derived from decaying marine life.

Core

The center of the planet earth is called the core and is made of dense iron metal under intense heat and pressure.

Coriolis force

The force created by the rotation of the planet on its axis and the resulting effect it has on liquid and gases like the ocean currents and the wind.

Corona

The sun's image in an eclipse.

Corrie lake

A glacially eroded depression high up on the side of a mountain that is filled with rainwater and post-glacial melt-water.

Corrie Lake

Corsica

A French island in the Mediterranean Sea and the birth place of Napoleon Bonaparte.

Cotton

A fibre used to make cloth and other products. China is the world's leading cotton producer, but cotton is an important source of employment in many southern U.S. states.

Costa Rica

A country in Central America between Nicaragua and Panama. The capital city is San Jose.

Crater

The feature that remains when a volcano erupts spewing out massive amounts of lava, ashes and rocks.

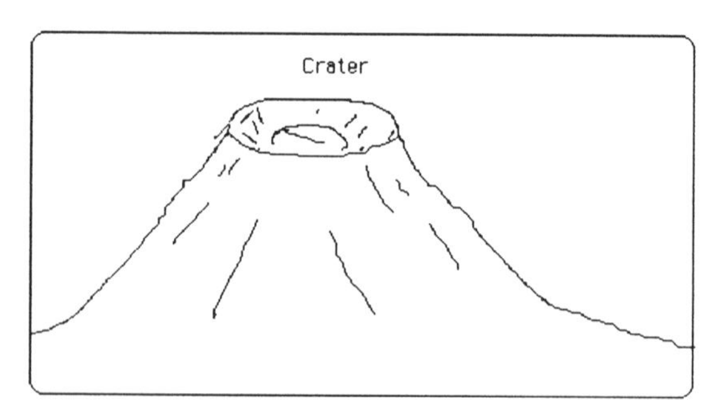

Fig. C.vi Crater

Crater Lake National Park

A wonderfully picturesque lake in an extinct crater in southern Oregon, which has been turned into a national park and attracts thousands of visitors each year.

Crete

A Greek island in the Mediterranean Sea renowned for its attractive beaches and amenable climate. It is an important tourist center because of its many historic connections with one of the world's oldest civilizations.

Crevasses

Deep fissures that abound on rivers of ice, known as glaciers, which are a result of differential movement as the glacier negotiates its way down the preglacial river valley into the lowlands below. There are longitudinal and transverse crevasses and often they are hidden beneath fresh falls of snow, where thin bridges hide the awning, gaping cavern below.

Crevasses

Croatia

A country of former Yugoslavia in southeastern Europe. Zagreb is the capital city.

Crust

The lithosphere, or crust of the earth is the solid, 'cooled down' outer layer of the planet that surrounds the asthenosphere. It is between 8 and 25 miles thick in places and on it ocean and soil are formed.

Cryoturbation

In periglacial conditions, a matrix of clay and pebbles is aligned to the melting water as it seeps through the soil turning the long axis of every stone in the same direction.

Cuba

An island country in the West Indies north of the Caribbean Sea. Havana is the capital city.

Cumulus

A fluffy accumulation of air, a cloud, that is usually white or dirty white and can mean fair weather when the clouds are high up and the sky is blue, but foul weather when they are dark and low and full of water vapor.

Currents, Ocean

The movement of warm and cold bodies of water on the surface of the earth. Their movements are influenced by the configuration of continental coastlines, the force and direction of the wind, and the Coriolis force.

Cyclones

A frontal weather pattern that brings heavy rain and storm conditions.

Cyprus

An island country and site of an ancient trade route in the Mediterranean Sea. It is a place of scenic beauty with hilltop castles, old churches, beaches and rugged mountains. The capital city is Nicosia.

Czech Republic

A country in central Europe. Prague is the capital city. Formerly, and together with Slovakia called Czechoslovakia.

Da Gama, Vasco

A Portuguese explorer, who opened up a sea route to India in 1498. He sailed around the Cape of Good Hope and north along the African coast to Malindi, from whence he headed east to Calicut.

Dalles

Deep river gorges caused by erosion, usually in slab basaltic rock.

Danube

A European river that rises in southern Germany and flows southeastwards into the Black Sea. Along its 1,700 miles it is harnessed by many dams to produce electricity and its navigable reaches are used for transporting goods.

Dardanelles

A strait that connects the Aegean with the Sea of Marmara. See also; the Black Sea.

Darjeeling

A city situated high on the slopes of the Himalayas in northern India. Its high altitude makes for a pleasant summer climate and many people move there to escape the dense heat of the regions lower down. It also gives its name to a local black tea that grows abundantly in the hillsides nearby.

Darling

A river (the longest) in southeastern Australia that flows through Queensland and New South Wales to its confluence with the Murray.

Day and Night

Half the planet earth can face the sun at any one time. This means that half the globe will be in daytime and half will be in night-time.

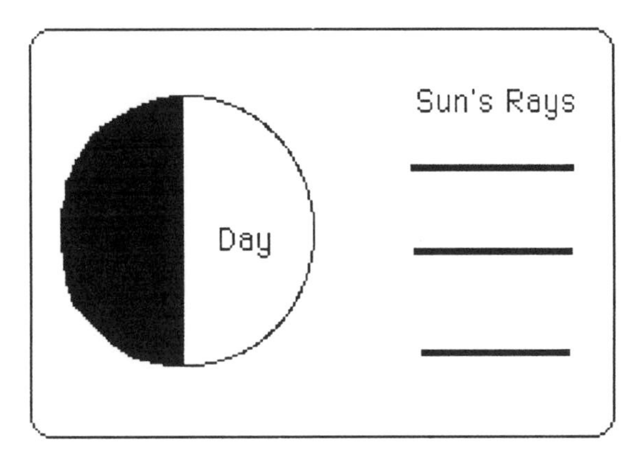

Fig. D iii Day & Night

Daylight Saving

A plan in which clocks are set one hour ahead of standard time for a certain period during the year to facilitate more daylight in the evening and conserve energy.

Dead Sea

A salt lake on part of the border between Israel and Jordan. At over 2,600 feet below sea level it is the lowest place on earth. The high rate of evaporation causes salt to build up in the sea and as a result swimmers find the water extremely buoyant.

Death Valley

A desert valley in eastern California and also the lowest elevation in the

United States, at 280 feet below sea level. It is a graben. Death Valley gets its name from the number of ill-fated, wagon train attempts at crossing it during the expansion to the western seaboard. Temperatures range in the 120s during the summer months.

Deforestation

When trees are systematically felled year, after year in a forest region, deforestation occurs, often with disastrous environmental impact. Clear-cutting is a particularly severe form of deforestation. Conversely, since trees are a renewable resource, afforestation techniques can correct this problem over time.

Delaware

A state in eastern U.S. Dover is the capital city.

Delta

The deposition that occurs when a river reaches the sea often creates a delta. Most large rivers enter the sea through deltas in their lower courses. The Nile and the Mississippi, for example, possess large delta regions where their load of detritus is dropped at the mouth of the river. Deltas eventually become fertile plains.

Demography

The study of human population, including its distribution and movement. Demographers study trends in age, sex, migration patterns, birth and death rates and other matters in relation to a country's standard of living and its population.

Dendritic

This is a drainage pattern called after the Greek word for tree - Dendron. The river system resembles that of the veins of a typical leaf where tributaries feed the main artery in a downward direction.

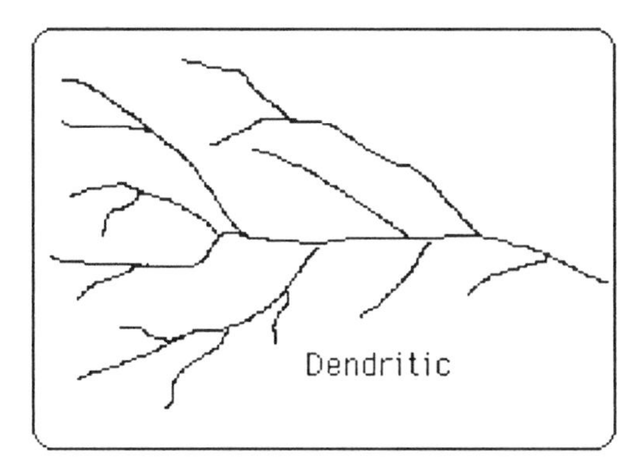

Fig. D i Dendritic Drainage

Denmark

A small country in northwestern Europe, occupying the Jutland peninsula and a number of adjacent islands in the Baltic and the North seas. It is renowned for its dairy produce and agricultural products in addition to porcelain and furniture manufacturing. Copenhagen is the capital city and is situated (mostly) on the island of Sealand between the Kattegat and the Baltic Sea.

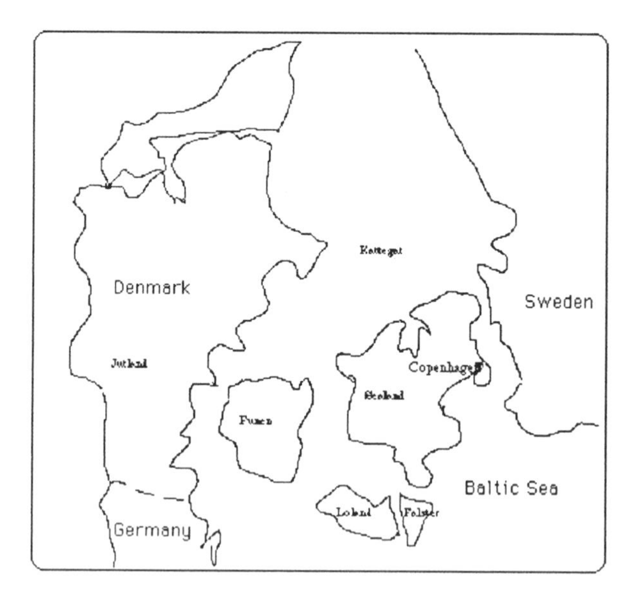

Fig. D ii Denmark

Denver

The capital city of Colorado also known as the "mile high" city. It is the gateway to the Rocky mountain region in terms of transportation, manufacturing and outdoor pursuits.

Deposition

The silt and mud that is dropped by the river when it slows down or stops. Likewise the detritus that a glacier dumps in its moraines when it melts.

Desalination

The process of extracting salt from water.

Desert

An arid region where there is an insufficient water supply and the soil is too dry to grow crops. However, many desert regions have their own flora and fauna suited and adapted to the arid conditions.

Detritus

Another word for the sediment that is carried along in a river or a glacier.

Detroit

The automobile capital of the world, Detroit city developed as a leading industrial center because of its strategic location on lake Erie and also because it was the home of Henry Ford, the Dodge brothers and Ransom Olds. The industry grew because there was a plentiful supply of labor and raw materials could be shipped inexpensively by lake. The city and its population is naturally marked by fluctuations in the automobile industry.

Dew

When warm air near the ground cools, it condenses until it forms a droplet of water. This is dew and is found on grass and leaves early on clear mornings.

Dew point

The point (temperature) where condensation causes water droplets to form.

Diamond

A hard element that is very valuable. Because of its value it is used for jewelry and because of its hardness it is used in industry.

Differential erosion

Different rock types are eroded at different rates due to differential erosion. Weaknesses are exploited faster than solidly cemented rocks.

Dinosaur

A prehistoric mammal that is thought to have become extinct as a result of either an ice age or because of the cataclysmic impact of a meteorite.

Distributaries

The streamlets and waterways that distribute water into the sea through a delta, from a river system.

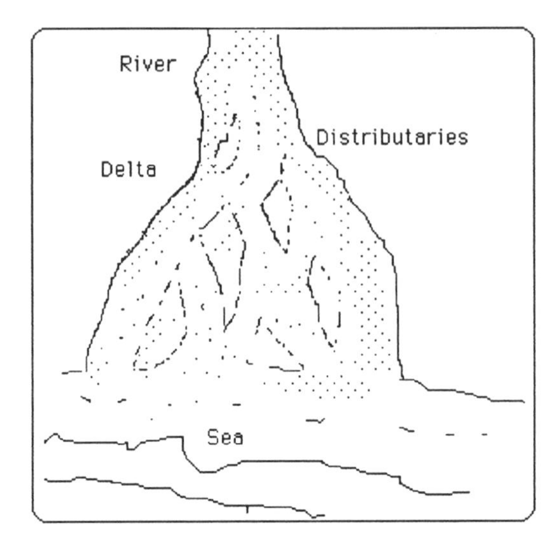

Fig. D iii Distributaries

Divide, (Watershed)

An upland region that separates river systems.

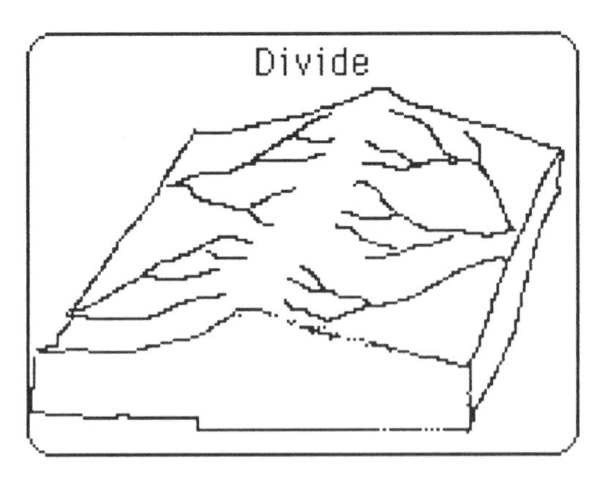

Fig. D iv Divide

Doldrums

A region in the mid-latitudes where the prevailing winds are often calm.

Dormant

When a volcano has not erupted for a couple of hundred years it is said to be dormant.

Douglas Fir

A softwood and the major source of timber in North America.

Dover, Straits of

The narrow channel between south-eastern England and northern France. It is rimmed to the north by spectacular white chalk cliffs rising above the ocean beaches to the windswept downs. A tunnel, beneath the channel, is being constructed jointly by the French and English to connect the island of Britain to the mainland of Europe.

Drainage Basin

Another term for catchment area. It denotes the area drained by a river and its tributaries.

Drowned river valleys

In post-glacial topography coastal areas were drowned when the ice melted and returned to the sea during a warm spell. Geographers estimate that the ice age sea-level was roughly 300 feet below that of present times. Rivers that cut channels to that sea were naturally drowned when the sea level rose again in the aftermath of the continental ice sheets.

Drumlins

Post glacial topography often displays drumlin formations. They are also known as 'eggs-in-a-basket' terrain. Deposits of boulder-clay laid down and sculpted by the moving ice flow result in these mounds of earth whose long axes indicate the direction of the ice movement.

Dublin

The capital city and cultural center of the Republic of Ireland.

Dune

A formation of sand that is a result of the prevailing action of the wind on the sea shore or in a desert region.

Dust Bowl

A major agricultural disaster in the mid and southwestern states during the 1930's. Erosion and high winds resulted in millions of tons of top soil being lost from the prairie states that had been ploughed and exposed to the elements.

Dyke

This is an intrusive igneous plug that invades a weakness in a rock. A dyke is typically in an upright position in the rock face, whereas its counterpart, a sill, is usually horizontal.

Earth

The planet that is third in order from the sun and upon which mankind dwells.

Earthquake

An upheaval of the earth's surface caused by tectonic plate movement.

Ebbing Tides

The condition of the tide as it returns to the sea.

Eclipse

The obscuring of one celestial body by another as in the case of the moon blocking the light and heat of the sun.

Ecology

The study of the relationships that living things have for each other and to their environment.

Ecosystem

A biological environment. All living things and their physical environment make up an ecosystem. It consists of communities and their natural surroundings such as climate, water and soil. Understanding how an ecosystem functions helps one know about the balance of nature.

Economic geography

The study of the production and distribution of goods and services for people and nations.

Ecuador

This is one of the smallest countries in South America, crossed by the equator from which it gets its name. The Andes mountain range covers much of Ecuador and agriculture predominates in the valleys where most of the people live. Quito is the capital city.

Edam

A small community in the Netherlands near Amsterdam renowned for a local brand of cheese of the same name.

Edinburgh

The capital city of Scotland located on the Firth of Forth.

Edmonton

The capital city of Alberta, Canada.

Egypt

An Arab nation in the northeastern corner of Africa. Most of Egypt is covered by arid lands and wind swept deserts. The multitude of inhabitants outside Cairo survive mainly on the results of irrigation from the nearby Nile river. Cairo is the capital city.

Eiffel Tower

A famous landmark in Paris that was designed by French engineer of the same name. He also designed the Statue of Liberty.

Einstein, Albert

One of the greatest scientists of all time,

who contributed much to our modern understanding of the universe.

Eire

The Gaelic name for Ireland. See also; Ireland.

Elevators

Transportation devices that, through their efficiency and speed helped change the look of our cities by fostering the development of skyscrapers.

El Nino

A warm current that periodically appears off the western coast of the Americas with resulting changes in weather patterns. Usually storms and drought can accompany an El Nino current and the climatic patterns for that year may not be in keeping with those of many previous years.

Ellipse

The path of a point that is not a correct circle.

Elliptical orbit of planets

The planets, including earth, move in

elliptical orbits about the sun. As a result of this revolution the earth is sometimes closer to the sun and at other times farther away.

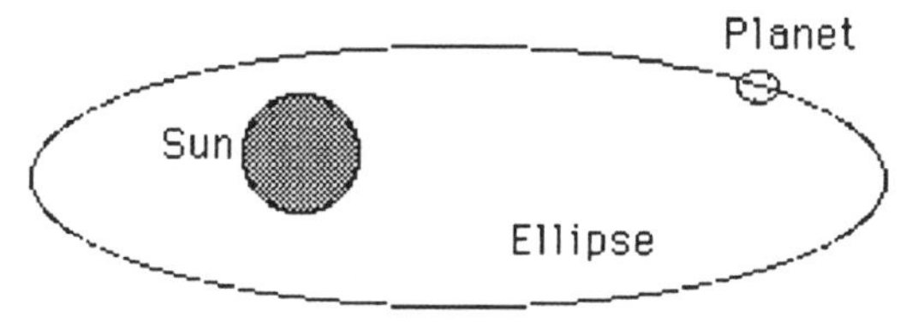

Fig. E.i Elliptical Orbit

El Salvador

A densely populated country on the western shores of Central America, bordering Guatemala and Honduras.

Energy

The ability to do work. Most of the energy on earth comes directly or indirectly from the sun. The sun's energy is stored in coal, oil and gas which we use as fossil fuels to keep the wheels of industry turning. Energy is captured from rivers and tides to turn turbines, from the wind to turn vanes, from the sun, from nuclear sources and from geothermal chambers.

England

The largest of the three countries that make up the island of Britain. Wales and Scotland border it to the west and north respectively. It refers to the land of the Angles as opposed to the Saxons, hence the name. London is the capital city.

Entrenched drainage system

When a river course experiences rapid down-cutting during a period of land-uplift, the features like meanders and cliffs are said to be entrenched.

Environment

The physical surroundings associated with living things.

Environmental determinism

The particular outlook in geographical understanding whereby the physical phenomenon is the most important for explaining man's dynamic relationship with his surroundings.

Environmental Protection Agency

An agency of the U.S. Government that establishes and enforces environmental standards and conducts research on the effects of pollution.

Eocene

The geologic era when mountain building was taking place. The Alps, Andes and Rockies were raised by tectonic movement during this time and mammals continued to proliferate.

Epicenter

The point on the surface of the ground directly above the focus of an earthquake.

Epoch

A subdivision of a geologic period of time, e.g. the Eocene epoch is part of the Tertiary period of the Cenozoic era.

Equator

An imaginary circle around the center of the earth equidistant from both the north and south poles.

Equatorial jungles

The lush vegetation of the equatorial region.

Equinox

The time, during the vernal and autumnal equinoctal positions, when the length of day and night are the same.

Era

One of the five main periods of geologic time marked by particular distinguishing features and events.

Eratosthenes

An early geographer who, around the year one hundred BC, painstakingly calculated the circumference of the earth.

Eric the Red

A Norwegian explorer who discovered and colonized Greenland. His son is said to have settled a colony on Newfoundland.

Erie

A lake in the Great Lakes chain on the path of the Niagara river. Also, a canal joining lake Erie to the Hudson River in New York and a city on the lake of the same name in Pennsylvania.

Fig. E ii Lake Erie

Erosion

When agents of the elements, for example rain, waves and wind wear-away, abrade or chemically attack rocks and soil, erosion is said to have taken place.

Erratic boulders

When large boulders are transported by ice or water or wind to a place where they are obviously not indigenous, they are deemed to be erratic.

Eruption

Active volcanos spew lava, ash and rocks in a typical eruption.

Escarpment

An anticline with one side decidedly steeper than the other.

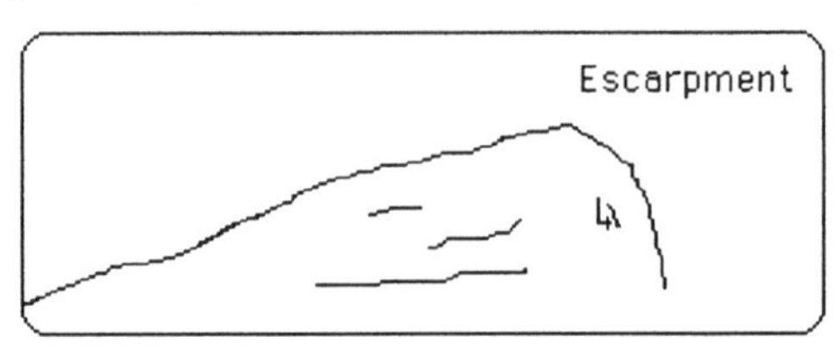

Fig.E.iii Escarpment

Esker

A subglacial river course results in a peculiar sand and gravel ridge formation that snakes across the land when the ice has retreated or melted.

Eskimo

A native people who dwell in the Arctic reaches of the globe, including Alaska, Northern Canada and Greenland.

Estonia

A country of northern Europe on the Baltic Sea. The capital city is Tallinn.

Estuary

An inlet of the sea or a wide mouth of a river where salt water meets with fresh water.

Ethiopia

A country in northeastern Africa bordering the Red Sea. The capital city is Addis Ababa.

Etna, Mount

A volcanic mountain in Sicily.

Euphrates River

A river that rises in eastern Turkey and flows into the Persian Gulf. Together with the Tigris, the Euphrates' flood plain is site of the oldest civilization (Sumerian, 3500 BC) in the world. The ruins of Babylon can be explored along the banks of the Euphrates.

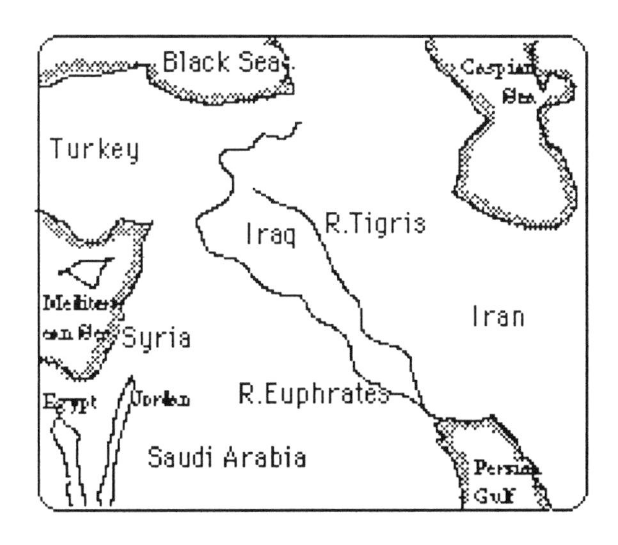

Fig. E iv Euphrates River

Europe

The continent that lies in the northern hemisphere between the Atlantic and Asia.

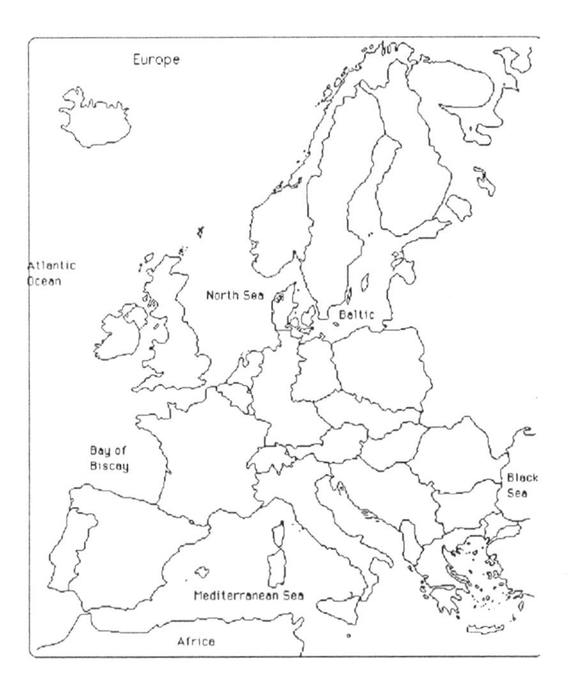

Fig. E v Europe

European Community

The common market formed in the late 50's by some European countries to unite their economic destinies and achieve a better focus in world trade.

Evaporation

When a liquid changes into a vapor, evaporation is said to have occurred.

Evening

The end of the day or the early part of the night, sometimes called twilight.

Everest, Mount

At 29,028 feet above sea level this is the highest mountain in the world. Mount Everest is situated on the border of Tibet and Nepal in the Himalayan range. It was climbed successfully in 1953 by Edmund Hillary and Sherpa Tensing.

Everglades

A large tract of swamp land in southern Florida, which is a flourishing natural habitat for all kinds of wildlife. It is located near lake Okeechobee.

Evergreen

Plants and trees that keep their foliage through-out the year, as opposed to deciduous trees that shed their's in winter.

Evolution

A gradual progressive change over time from primitive to advanced society.

Exfoliation

When a rock like granite is weathered, especially with freeze-thaw action, exfoliation occurs. The rock expands under intense solar radiation by day and contracts under extreme cold at night. These differential temperatures cause the structure to collapse and it crumbles, or peels like an onion.

Exotic flora

Vegetation that is colorful and different to the native vegetation.

Exploration

Discovery and expeditions to find out about new places.

Exports

Goods sent out of a region or country in exchange for other goods, services or money.

Extinct - volcano

A volcano that has not erupted for many hundreds of years.

Extinction

Some plants and animals have been subjected to pressures by the modernization process of today's world and have been driven to extinction. Other species died out because of changes in climate which brought about glacial periods, and by cataclysmic collisions with asteroids.

Extrusive Igneous Rock

Volcanic lava that is forced through the vent onto the surface and which cools quickly resulting in small crystals. Basalt is an example of this kind of cooled lava.

Eye - hurricane

As a hurricane passes over, there is a period of calm in the center when it may appear that the storm is past, but it really means that the second half of the battering is about to begin. This is a result of the intense cyclonic revolution of the winds in the storm about the center.

Factory

A building in which products are manufactured.

Faeroe Islands

A group of Danish islands in the north Atlantic between Iceland and the Shetland islands.

Falkland Islands

A group of islands off the tip of South America east of the Strait of Magellan. The islands are a British dependency, although Argentina claims them. The capital city is Stanley, and this is where most of the 2,000 inhabitants live.

Fall Line

A geomorphological feature where a series of waterfalls and rapids occur due to differential erosion. The waterfalls and rapids associated with fall lines can be harnessed to generate electricity. A fall line exists on the eastern coast of the U.S. between New York state and Alabama. Along this line numerous rivers are harnessed in hydro-electric schemes, including the Potomac, the Susquehanna, the James, the Roanoke, the Savannah, and the Chattahoochee.

Fallout

In the event of a nuclear explosion the radio active particles that descend to earth are called collectively *fallout*.

Fallow

Land that is ploughed but not seeded for one or more seasons in order to kill weeds and make the soil richer.

Famine

An acute and prolonged general scarcity of food, causing sickness and death.

Farming

The cultivation of the land for food production.

Fault line

A geologic weakness in a rock structure. Usually a fracture results in a side slipping so that it appears displaced from its original position.

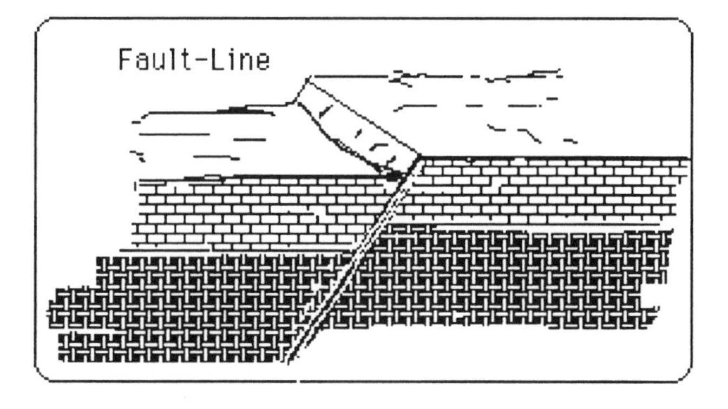

Fig. Fi Fault Line

Fauna

The animal life of a particular region, for example, the fauna of Texas.

Fertile Crescent

The rich farmland encompassing the Tigris and Euphrates flood plains where the first civilization prospered more than 5000 years ago. See also; Euphrates.

Fertilize

A process of adding natural or chemical constituents to the soil to enrich it and make it more productive.

Fetch

This is the name given to the expanse of ocean over which the wind blows and causes wave activity.

Field-work

Experimental practical activity undertaken by students alone or in teams as part of outdoor geography.

Fiji

An island chain in the southwest Pacific. The capital city, Suva, is on one of the islands (Viti Levu). This island, like most of the other eight hundred or so, is a volcanic cone surrounded by corral reefs.

Finger lake

A long narrow glacial lake. Also the name given to a group of glacially scoured or dammed lakes in New York state.

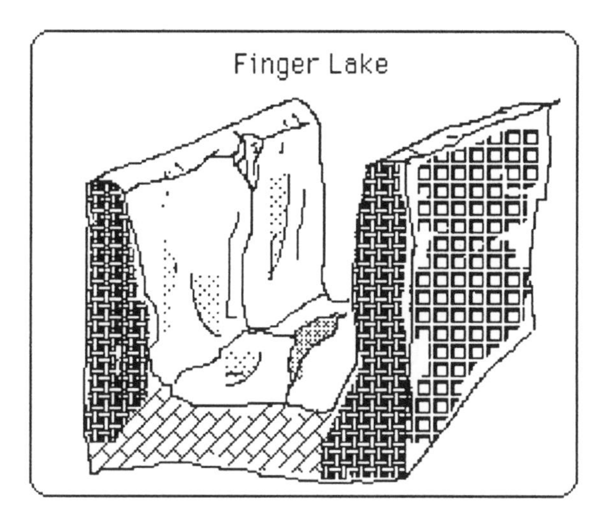

Fig. Fii Finger Lake

Finland

A country in northern Europe bordering on the Gulf of Bothnia and the Gulf of Finland. The capital city is Helsinki.

Fiord/Fjord

A long narrow inlet of the sea bounded by steep cliffs. A direct result of glacial activity where a u-shaped valley is drowned. Fiords usually have rock barriers at their mouths known as skerries. Norway has a coastline of numerous spectacular fiords.

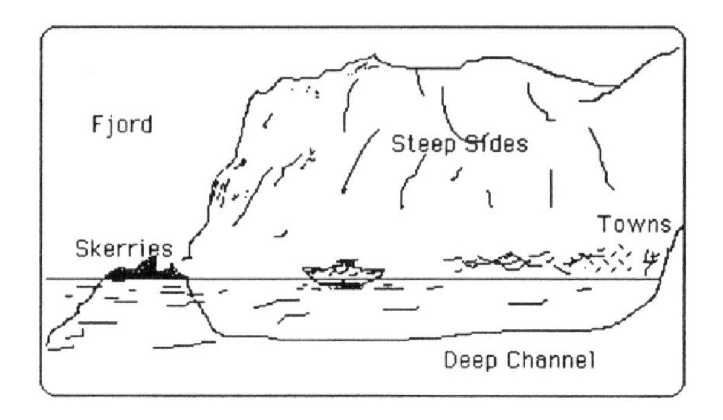

Fig. Fiii Fjord

Fir

Cone bearing evergreen tree with needles. Fir trees produce a robust working wood for carpentry.

Fire break

A strip of land cleared to stop the spread of fire in a forest zone.

Firn

The old uncompacted snow that accumulates on the top of a glacier.

Fish

The aquatic cold blooded vertebrate animal that lives in water and is a source of food for man.

Fish Ladder

An ascending series of pools so arranged as to allow fish to leap from one to the next and pass over dams and waterfalls to reach their spawning grounds higher up.

Fishing Industry

The catching, marketing, processing and conservation of fish for commercial advantage.

Fission

The splitting of the atom in a nuclear reaction to create enormous amounts of energy.

Fissure

A long narrow deep crack, usually in rock strata.

Flax

A plant that is grown to produce fibers and seeds. The fiber plant is used to produce high quality paper, rope and thread. The flax seed is used to make linseed oil for paints and varnishes. France, Romania, Poland and other European countries grow fiber flax.

Canada, Argentina, India and the Soviet Union grow flax seed.

Flint

A hard, sharp rock that was used by early man to ignite kindling and start a fire. It was also a useful scraper and tool-head since it is hard and sharp and easily chipped.

Flood plain

The plain along a river formed by sediments deposited by floods. A low-lying area near the channel that is liable to annual flooding.

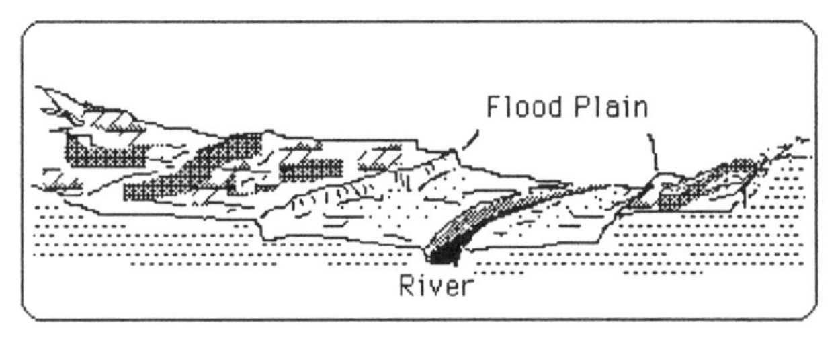

Fig. Fiv Flood Plain

Flooding

After severe rainfall, or melting of the snow pack, swollen rivers burst their banks and levees, and inundate the surrounding land. This results in flooding to low-lying areas and flood-plains.

Flora

The plant life of a specified region, for example, the flora of Texas.

Florida

A state in the southeastern U.S. extending into the Gulf of Mexico. The capital city is Tallahassee.

Florida Keys

A chain of small islands extending southward from the tip of Florida.

Fluorocarbon

A non-reactive synthetic organic compound containing carbon and fluorine used in aerosols and some lubricants. These elements have been found to be damaging to our atmosphere.

Focus (Earthquake)

The starting point of an earthquake, usually deep within the lithosphere, and directly under the epicenter.

Fog

A thick obscuring mist, a water vapor in the atmosphere. Fog typically forms when cold air mixes with warm air.

Fold Mountain

Rock layers folded or bent by pressure caused by earthquakes or tectonic movement.

Food Chain

A sequence of organisms in a community in which each member of the chain feeds on the member below it, for example, fox, rabbit, grass.

Food production

The growing and manufacturing of food for human consumption.

Forage

Food provision for domestic animals, for example, hay.

Forest

A thick growth of trees and underbrush covering an extensive tract of land.

Forest products

Materials that are produced from forest industries include fuel, building material and paper.

Fossil

Traces of plant or animal life of some previous geological period, preserved in rock formations in the earth's crust.

Fossil fuel

Coal, petroleum and natural gas are dug from the earth and are known as fossil fuels. They are non-renewable and when used up, they are gone.

France

A country in western Europe whose coastline borders the Atlantic, the North sea and the Mediterranean. The capital city is Paris.

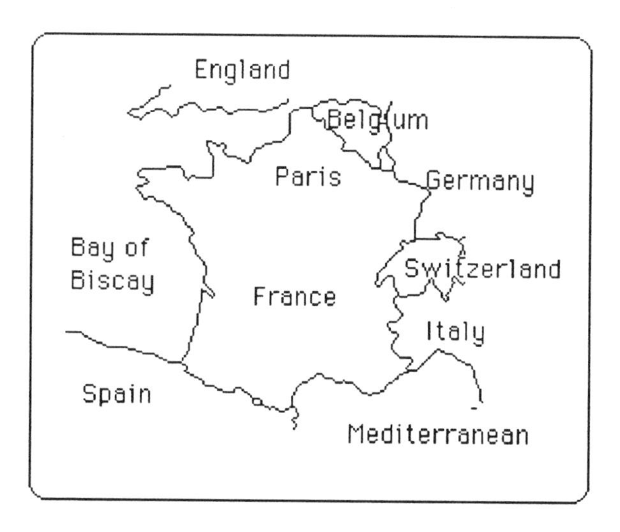

Fig. Fv France

Freeze/thaw action

The kind of shattering erosion that takes place in rock structures when water, lodged in fissures is exposed to conditions of freezing and thawing. The freezing causes the water to become ice and expand by 10 percent. Constant freezing and thawing causes the rock to exfoliate or shatter.

Freezing point

That point when water turns to ice.

French Alps

The high mountain range on the borders of France, Switzerland and Italy.

Fresh water

Water that does not contain salt.

Front

A weather condition caused by the mixing of bodies of warm and cold air.

Frost action

The name given to the destructive forces of ice as rocks are weathered.

Fuji-yama (Mount Fuji)

The highest mountain (12,388 feet) in Japan, on the island of Honshu. It is an extinct, cone-shaped volcano that is snow-capped throughout most of the year. Thousands of Japanese climb to its summit each year.

Fusion

A nuclear reaction that converts atoms into enormous quantities of energy.

Galapagos Islands

A group of volcanic islands to the west of Ecuador, to which they belong, named after the giant turtles that inhabit them. Other interesting and unique birds, mammals and animals live on these islands also. The study of these species was partly instrumental in leading Charles Darwin to theorize about evolution in the last century.

Galaxy

A system of stars, star clusters and interstellar matter that constitute the universe. Galaxies are either spiral shaped or elliptical and can be millions of light years across. Our solar system lies at the edge of a spiral galaxy called the Milky Way.

Galena

A bluish grey mineral with metallic lustre, that makes up the principal ore of lead.

Galileo

An Italian scientist, geographer and mathematician who opened up vast new fields of knowledge towards the middle of the 18th century.

Gamma ray

A photon emitted spontaneously by a radioactive substance.

Ganges River

A river that rises in the Himalayas and flows southeast to its confluence with the Brahmaputra and eventually spills into the Bay of Bengal through the vast Ganges delta. The river is central to the Hindu way of life and is considered sacred.

Gas, Natural

A fuel that was formed, like oil, beneath the earth's surface millions of years ago. It is mined just like oil is and is used to heat homes, in industry and as a raw material for other substances.

Gaucho

A cowboy/cowhand of the South American Pampas.

Gaza

City and coastal strip along the Mediterranean in Palestine. A disputed political unit, it has been administered by Egypt and by Israel at different times.

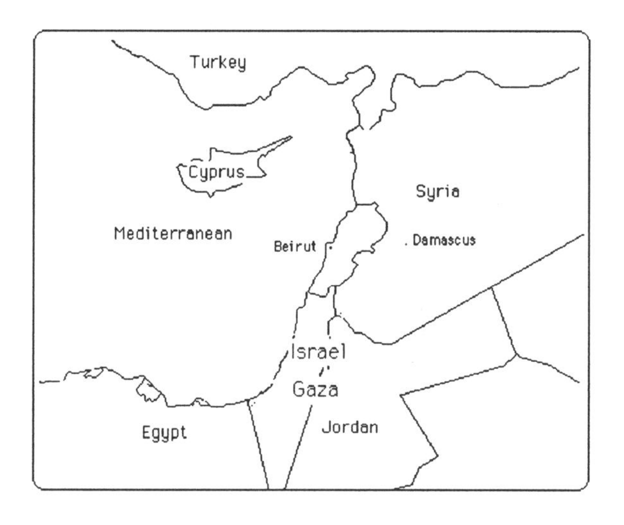

Fig. G i Gaza

General Agreement on Tariffs and Trade

An international treaty that provides a forum, at the Geneva headquarters, for agreements on trade barriers and trade related disputes.

General reference map

The ordinary street map that we use every day.

Geneva

A canton and its capital city in Switzerland on the river Rhone. Also a lake on the river Rhone bordering Switzerland and France.

Geo-thermal energy

A source of energy that utilizes the earth's interior heat.

Geocentric Theory

The Ptolemaic view of the universe with the earth as the center, as opposed to the heliocentric view expounded correctly by Copernicus.

Geodesic

The shortest distance between two places on the globe. Also known as a great circle route. To calculate a geodesic between Seattle and London, simply place a string on the globe with an end at each of the two cities. Note that this is not necessarily a straight line on a map.

Geography

The study of man in relation to his environment.

Geological Survey Map

A topographical depiction that is carefully designed to assist geologists carry out surveys.

Geologist

A scientist and geographer who studies rock strata and composition.

Geology

The study of the history of the earth and its life as recorded in rocks.

Geomorphology

The study, that seeks a genetic interpretation of the features of the earth's surface. Geomorphologists study the origin and arrangement of the earth's features.

Georgia

A state in southeastern U.S. The capital city is Atlanta. Also a nation bordering on the Black Sea. The capital city is Tbilisi.

Geostrophic

The deflective forces caused by the rotation of the earth. The geostrophic wind circles high above the surface unimpeded by windbreaks or other obstacles.

Germany

A country in central Europe between the North and Baltic seas. After the Second World War it was partitioned into West and East Germany. In 1991, East Germany became free of Russian dominance and Germany was re-united. The capital city is Berlin.

Geyser

A spring of volcanic origin that spouts forth intermittent sprays of water and steam.

Ghana

A tropical country in west Africa. The capital city is Accra. Because of the tropical climate, hardwoods are produced and exported. Cacao seeds, which are used for making chocolate, are also grown and exported.

Gibraltar

A British dependency occupying a narrow peninsula on the southern coast of Spain. Strategically, Gibraltar controls a vantage point on the entrance to the Mediterranean Sea and as a result it has always been a particularly suitable site for military purposes. The spectacular rock of Gibraltar is a massive limestone block jutting into the Mediterranean.

Glaciated U-shaped valley

A pre-glacial river valley that has been widened and deepened by a glacier and displays a characteristic u-shape with truncated spurs. Contrast with a typical river valley which is V-shaped with interlocking spurs.

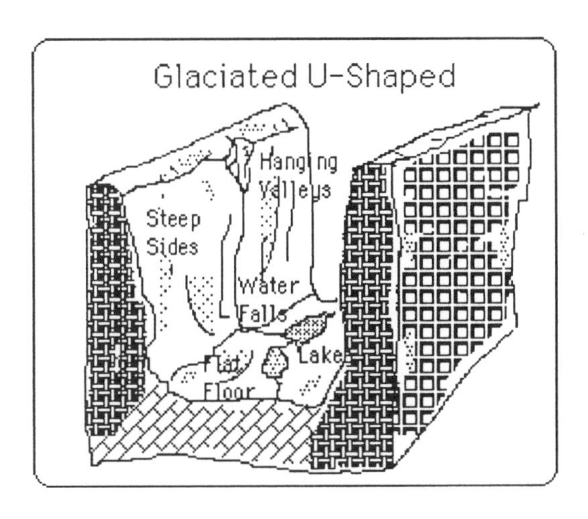

Fig. G ii Glaciated U-Shaped Valley

Glacier

Often called a river of ice, it is a large mass of snow and ice that forms in areas where the rate of snowfall constantly exceeds the rate at which the snow melts. It moves slowly outward from the center of accumulation, usually down a valley until it reaches the sea and/or warmer temperatures where it melts. Sometimes it breaks off into the sea to form icebergs.

Global destruction

The catastrophic results that might ensue following a nuclear disaster, since radioactive fallout, and other poisonous emissions into the atmosphere would result in major destruction to life on earth. Other factors including global warming and ozone layer depletion, could hasten the destruction of our planet.

Global stability

The ideal that in the global village we must all work together to improve our lot and help one another along.

Global warming

The hot house effect is created by the

burning of fossil fuels and the emissions of exhaust and other pollutants from our cities, homes and factories.

Globe

A spherical representation of planet earth that is true to scale in terms of shape, area and direction.

Gneiss

A metamorphic rock corresponding in composition to granite.

Gobi Desert

A desert in Mongolia, in east central Asia.

Gold

A malleable, ductile, yellow metallic element that is very valuable and is used in jewelry, coins and dentures.

Golden Gate

A strait in western California connecting San Francisco Bay with the Pacific.

Gorge

A narrow steep-walled canyon.

Graben

A rift valley.

Gradient

Ascending or descending with a uniform slope.

Grand Banks

A major fishing ground off Newfoundland. Fish are drawn here to feed on the plankton that abounds in the shallow waters of the continental shelf.

Grand Canyon

A deep gorge on the Colorado river, Arizona. It is a major tourist attraction because of its spectacular geographic formations. The entrenched drainage pattern is the result of down-cutting at the same time as isostatic recoil was taking place.

Grand Coulee Dam

The greatest hydro-electric power supply in the U.S. It is built on the Columbia River in eastern Washington. Coulees are remnants of glacial over-flow channels and can easily be dammed to harness hydro-electric schemes.

Granite

A very hard, crystalline rock of igneous origin, formed well below the surface of the earth by intense heat. It is usually gray to pink in color, and consists of feldspar, quartz and mica.

Graphic scale

A calibrated linear diagram on the corner of a map that is drawn to depict the scale showing the inches on one side and the miles on the other.

Gravel

A loose mixture of pebbles and rock fragments coarser than sand, usually found on the shores of rivers and lakes.

Gravitation

The force by which every particle of matter attracts and is attracted by every other particle of matter.

Gravity

The force that draws every body in the earth's sphere towards the center of the earth.

Great Australian Desert

The central arid regions of Australia.

Great Barrier Reef

The corral reef off the Queensland coast of eastern Australia. It is the largest group of corral reefs in the world and a haven for divers. See also; Barrier Reef.

Great circle

A geodesic, or the shortest distance between two places on the globe. It is formed by drawing a circle on the surface of the sphere whose plane passes through the center. See also; Geodesic.

Great Lakes

A post-glacial chain of fresh water lakes in northeastern U.S. that spills into the St. Lawrence seaway. The initial letters of the five lakes spell HOMES: Huron, Ontario, Michigan, Erie and Superior.

Great Wall of China

A stone and earth wall extending across northern China built as a defence against invaders in the 3rd century B.C. It is 15 - 30 feet high, 12 - 30 feet wide and 1,500 miles long.

Greece

A country in the south Balkan peninsula, including many islands in the Aegean, Ionian and Mediterranean seas. The capital city is Athens.

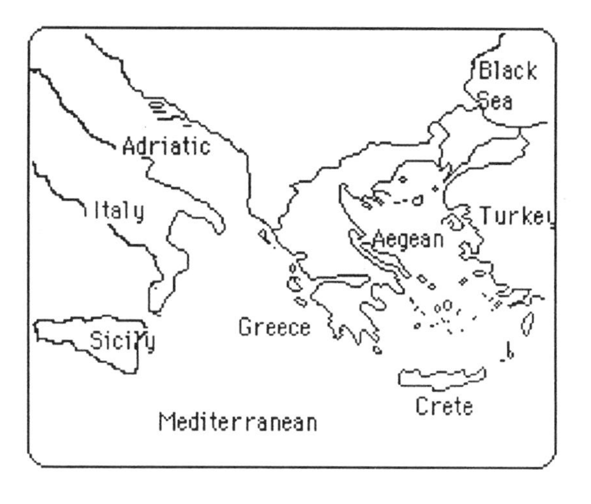

Fig G. iii Greece

Green Belt

A designated zone that surrounds a metropolitan area that is set aside for park-lands and greenery.

Green house effect

The build-up of heat associated with the retention of solar radiation and heat at the earth's surface by carbon dioxide accumulations in the earth's atmosphere.

Green Revolution

A progressive movement founded to establish more and improved food supplies for the impoverished regions of our earth. The development of new varieties of food plants, together with improved agricultural techniques, result in greatly increased crop yields.

Greenland

The world's largest island, owned by Denmark, situated off the coast of northeast North America. Because of its northerly latitude, it is mostly covered by ice.

Greenwich Mean Time

The basis for standard time derived from the central meridian that passes through the ancient village of Greenwich, England, the site of the oldest observatory, which is still in use today.

Gregorian Calender

A corrected form of the Julian calender, introduced by Pope Gregory XIII in 1852, and now used in most countries of the world. It provides for a year consisting of 365 days and one extra in leap years.

Grenada

The southernmost island of the windward group of the West Indies. The chief export is nutmeg and other spices. The capital city is St. George's.

Grid-lock

A phenomenon that can theoretically occur in overcrowded cities at rush hour, when all the traffic snags and nobody moves.

Grizzly bear

A large, ferocious, brownish bear of North America, having a shoulder hump and long front claws.

Gross National Product

The total value of a nation's annual output of goods and services, produced during a given time period. It is one of the methods of measuring a country's economic performance.

Groundhog Day

On February 2nd according to an old tradition the groundhog comes out of hibernation. However, if it is sunny and he sees his shadow he supposedly returns to his hole for 6 more weeks of winter weather.

Ground water

Fresh water found underground in porous rock strata and soils. Ground water sometimes comes to the surface as a spring.

Guam

The largest of the Mariana Islands in the western Pacific. It is a territory of the U.S. The capital city is Agana.

Guatemala

The most populated Central American country. The capital city is Guatemala City.

Gulf

A large area of ocean, larger than a bay, reaching into the land. e.g. Gulf of Alaska, Gulf of Bothnia, Gulf of Mexico.

Gulf of Mexico

A large extension of the Atlantic, surrounded by the U.S. and Mexico. It is the location of some of North America's richest offshore oil deposits.

Gulf Stream

A warm ocean current about fifty miles wide flowing from the Gulf of Mexico along the east coast of the United States and turning east at the Grand Banks towards Europe. It has a modifying effect on the climates of the countries in the northern latitudes, where it becomes the *North Atlantic Drift*.

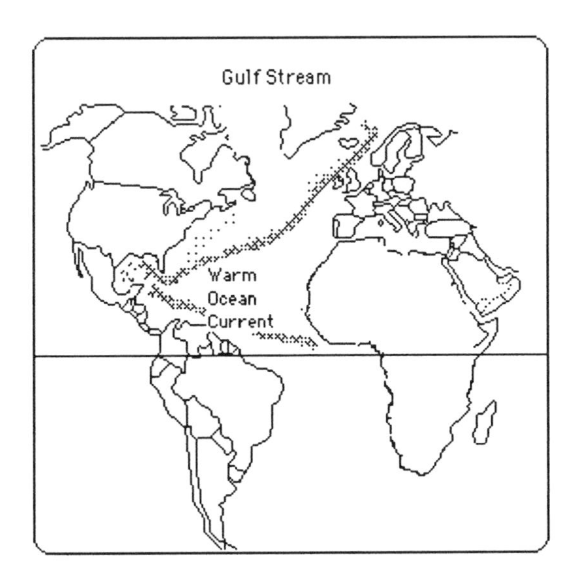

Fig. G iv Gulf Stream

Guyana

A county on the northern coast of South America. It was one of the first areas to be settled by Europeans. The capital city is Georgetown.

Guyot

A flat-topped steeply rising sea-mount, called after the U.S. geologist Arnold Guyot (1807-1884).

Habitat

The native environment where a plant or animal lives.

Hadrian's Wall

A stone wall across northern England, built by Hadrian in the second century to keep the northern tribes at bay.

Hail

Frozen rain that falls as a form of precipitation.

Haiti

A country in the West Indies on Hispaniola. Port au Prince is the capital city.

Half-life

The period required for the disintegration of half of the atoms in a sample of some specific radioactive substance.

Halibut

A large edible flat fish found in northern seas.

Halifax

Capital city and port of Nova Scotia in Canada.

Halley's comet

A celestial body named after the man who predicted its periodic appearance every 77 years or so. It was last seen in 1986 before it headed off on its elongated elliptical orbit that will bring it back again in the year 2063.

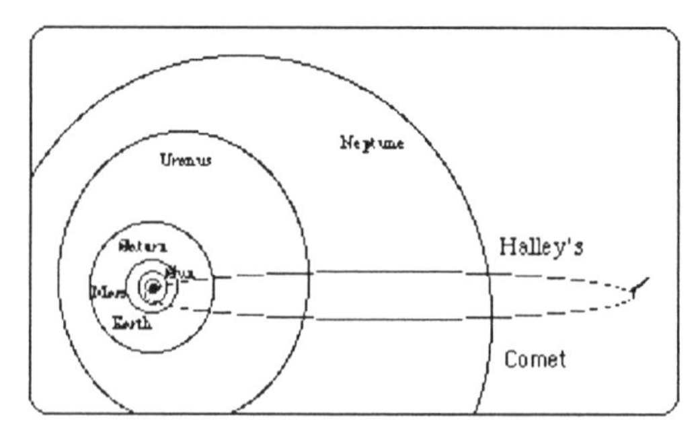

Fig. H i Halley's Comet

Hamburg

A city and seaport in northern Germany, on the Elbe river. It gave its name to the Hamburg Steak or hamburger.

Hanging valley

A post glacial, often very picturesque, waterfall that cascades into a u-shaped valley from a lake high up on the side of a mountain.

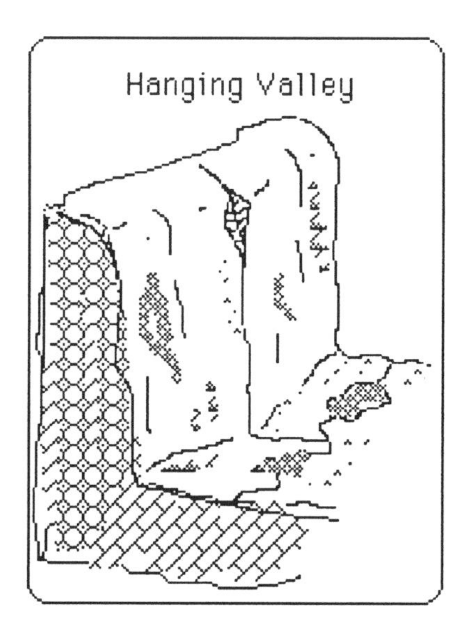

Fig. H ii Hanging Valley

Harbor

A protected inlet or arm of the sea where vessels can find safe moorage.

Harlem

A section of New York City in upper Manhattan bordering on the Harlem and East rivers.

Harrow

A heavy frame with spikes or sharp edged disks drawn by tractor and used for breaking up and leveling ploughed ground, covering seeds, rooting up weeds and so on.

Harvest

The time of the year when matured grain, fruit, vegetables and other agricultural produce are reaped and gathered in.

Harvest Moon

The full moon at or near the autumnal equinox on September 22nd or 23rd. It often appears reddish and larger since it is low on the horizon.

Havana

Capital city and port of Cuba in the Gulf of Mexico.

Hawaii

A group of tropical islands in the north Pacific, whose native inhabitants are of Polynesian descent. It is the only island state in the U.S. Honolulu is the capital city, situated on Oahu island.

Hay

A grass, alfalfa, clover and so on cut and dried for use as a fodder crop for livestock.

Hazardous waste

The waste from nuclear plants and toxic effluent that may damage human health or pollute the environment if not properly treated and disposed.

Headland

A point of land that juts out into the ocean.

Headland

Head waters

The small streams that are the sources of a river.

Headward erosion

This is a form of migration of the river upstream as the river eats into the upland behind it.

Hebrides

A series of islands off the western coast of Scotland in the Atlantic.

Heliocentric Theory of the Universe

The theory expounded by Copernicus that the sun is at the center and the other planets revolve about it. Note that it is opposed to the Ptolemaic geocentric theory of cosmology.

Helsinki

The capital city and port of Finland on the Gulf of Finland.

Hemisphere

The earth is divided at the equator into the northern and southern hemispheres. Hemi means half and sphere means circle, ball or globe.

Hemlock

North American evergreen tree of the pine family with drooping branches and short needles. The wood is used for carpentry.

High pressure

This term connotes a period of stable air when the barometric pressure is high. It is usually characterized by blue skies, white cumulus clouds and sunshine.

Hill

A raised part of the earth's surface, lower than a mountain.

Himalayan mountains

A mountain range in south central Asia extending along the India/Tibet border. The highest peak is Mt. Everest, which at 29,028 feet is the highest in the world.

Holland

A southern province of the Netherlands in northwestern Europe.

Honduras

A country in Central America. The capital city is Tegucigalpa. Many Hondurans make a living by growing coffee beans or by harvesting bananas that grow in this tropical climate.

Hong Kong

A crowded peninsula ruled by Britain but attached to the southern coast of China. In the past Hong Kong served as a port for British trade with China, but the end of British rule will occur in the year 1997. The capital city is Victoria.

Honolulu

The capital city of Hawaii on Oahu.

Honshu

The chief island of the Japan group.

Horizon

As far as the eye can see. The line in the distance between the sea and the sky, or where the two seem to meet.

Horizon (soil)

A separate layer of soil usually distinguishable by color. Ideally, humus is the top layer with a rich dark color known as the A horizon. This is topsoil and is the best layer in which to grow plants and crops. On top of the bedrock is usually a layer of partly weathered rock and this is called the C horizon. Between these two layers, a third stratum may form over tens-of-thousands of years, called the B horizon. Rainwater percolates through the humus and dissolves some minerals which are washed down and accumulate in the B horizon to form an iron pan. This often forms a hard layer that is impervious and traps water resulting in poor drainage and the growth of bogs and marshes. It is necessary to deep-plough these areas to revitalize the soil again and to re-introduce the minerals onto the surface layer.

Horse Latitudes

A region of calm centered on the Tropics of Cancer and Capricorn.

Horseshoe Falls

A waterfall, part of Niagara Falls.

Horn, Cape

A headland forming the southernmost point of South America on Tierra del Fuego.

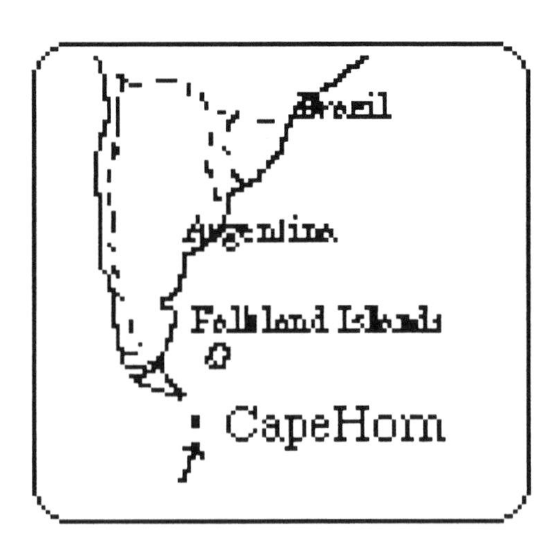

Fig. H iii Cape Horn

Horn of Africa

The easternmost projection of Africa into the Indian Ocean at Somalia, Ethiopia and Djibouti.

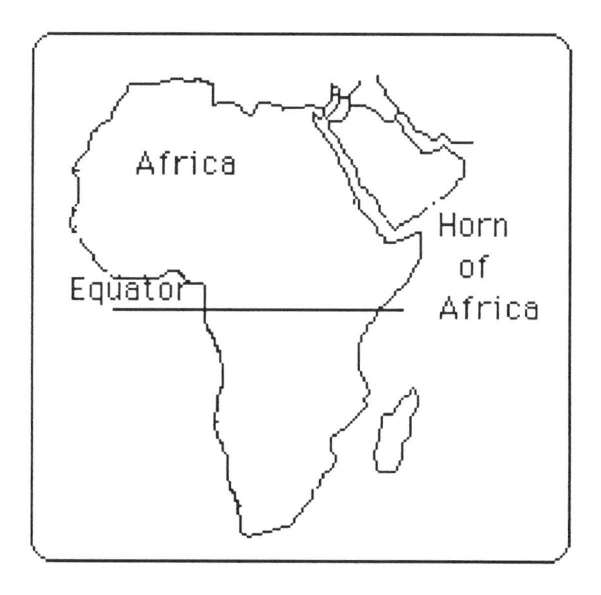

Fig. H iv Horn of Africa

Hot spring

This is a spring where the water is heated by geo-thermal means. Either contact directly with the magma chamber or hot gases from within the earth heat the water and often these springs are rich in minerals.

Hudson Bay

A large sea inlet in northern Canada named after the English explorer.

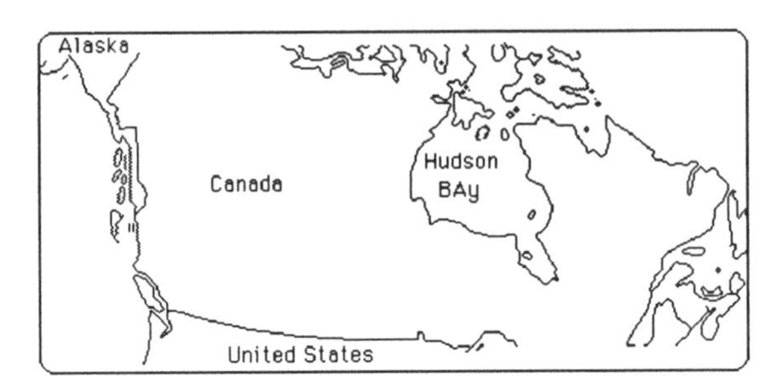

Fig. H v Hudson Bay

Hudson river

A river that rises in the Adirondack mountains and flows south into New York bay.

Humidity

The moisture in the air is called humidity.

Humus

The rich dark topsoil in the A horizon that is suitable for growing crops. See also; Horizon.

Hungary

A country in central Europe. The capital city is Budapest.

Hurricane

A large and powerful storm with strong winds and heavy rain. These conditions develop when warm air is accompanied by a source of moisture, like a warm ocean. The air rises rapidly and condenses. Then the Coriolis force causes the storm to be deflected in an anti-clockwise motion in the northern hemisphere. In the center there is an "eye" of calm air about which the storm clouds swirl at speeds reaching up to one hundred miles per hour. It is also called a typhoon in the western Pacific.

Hydraulic action

The erosive power of the river or the sea caused by its weight and the friction of flowing water.

Hydro-carbons

Another term for the fossil fuels, which are petroleum, coal and natural gas. Decaying remains of plants and animals on the sea floor are covered by sediment and over millions of years are converted into hydrogen and carbon substances, hence hydro-carbons. These hydro-carbons have been a mainstay in our energy supply in the past fifty years, but in light of the collateral impact to the environment and their imminent disappearance, new methods of providing energy are being researched and promoted.

Hydro-electric stations

Water generated electricity is a major source of our energy supply. It is caused by water flowing downwards that turns a turbine and converts the energy into electricity.

Hydroponics

A scientific experiment at growing plants in water or gravel.

Hydrosphere

The water parts of the earth, including oceans, lakes, rivers and ground water.

Iberian Peninsula

The landmass to the south of France in Europe that contains Spain and Portugal.

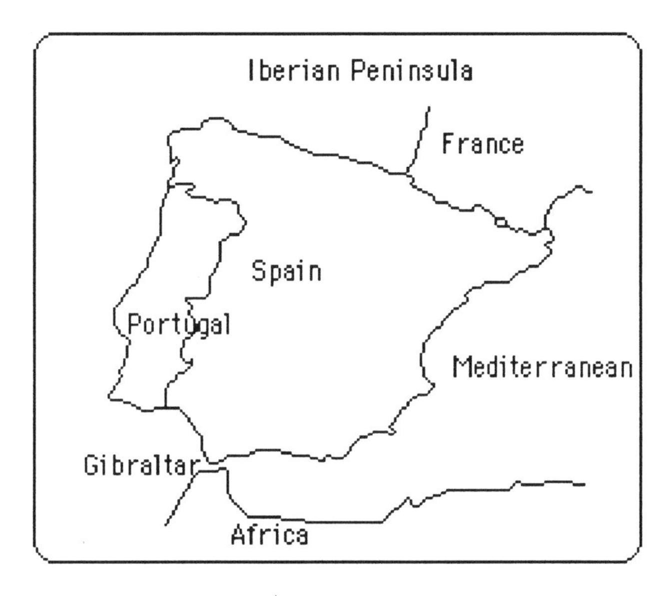

I.i Iberian Peninsula

Ice

The solid substance that results when water freezes. Water that freezes in pipes and rock fissures will expand by approximately 10% and will often cause them to burst.

Ice Age

A period in the history of the world when large ice sheets lay on top of much of the land. Water locked in these ice sheets originally came from the sea so that the sea level was lower during these times. The weight of the ice forced the land masses down so that the general topography was also lower at these times.

Ice falls

That part of a glacier by the snout where crevasses and seracs are in a precarious position and tumble down to the valley below.

Ice sheet

The large covering of ice that encompasses the polar regions of the globe. Ice sheets grow larger during ice ages so that they extend over the land and ocean towards the equator.

Iceberg

A large chunk of ice which falls off a glacier snout into the sea. It floats and is borne by the wind and tides in the prevailing direction. Icebergs are a major danger to shipping since they are up to ninety per cent submerged and can sink ships when collisions occur.

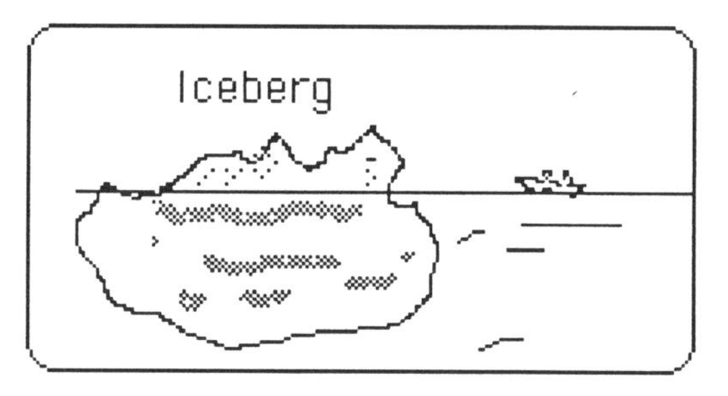

I.ii Iceberg

Iceland

An island of volcanoes, hot-springs and glaciers, in the North Atlantic. The Icelandic inhabitants colonize a thin coastal region around Reykjavik, the capital city.

Idaho

A state in the northwestern U.S. that is famous for the production of potatoes. The capital city is Boise. Idaho is one of the most beautiful "outdoor" states in the U.S. It has scenic snow-capped mountains that feed many rivers and rushing streams, placid lakes and picturesque forests. The rugged Hell's Canyon on the Snake river is deeper than the Grand Canyon.

Igloo

An Eskimo dwelling, made of blocks of snow that spiral upward in a dome shaped structure.

Igneous rock

Volcanic plutonic intrusions that cool down to form a crystalline rock. When magma cools granite is formed and when lava cools basalt is formed.

Illinois

A state in the central U.S. The capital city is Springfield. Also a river that is a tributary to the Mississippi. Chicago, on the shores of Lake Michigan, is the third largest city in the U.S. Because of its strategic location on the Great Lakes and in the Mid West, Chicago is the hub of transportation by air, road, rail and water.

Immigrant

A person who relocates for personal reasons to another country and makes it his adopted home.

Inca

A rich empire of South American Indians that prospered in the hinterland of Peru, before the Spanish conquest.

India

A country situated south of the Himalayas between the Bay of Bengal and the Arabian Sea. New Delhi is the capital city. India is one of the most populated countries of the world and the majority of its inhabitants have a poor standard of living.

Indian, American

Another name for Native Americans who pre-dated the European settlers.

Indian Ocean

The large body of water off the coast of East Africa and extending as far as the western shores of India.

Indian reservation

A tract of land set aside by the Government for Native Americans when white settlers colonized the United States.

Indiana

A state in central U.S. Indianapolis is the capital city.

Indochina

The eastern half of a peninsula in southeast Asia extending into the South China Sea. It comprises the countries of Kampuchea, Laos and Vietnam.

Indonesia

A country in southeast Asia comprising Sumatra, Java, Borneo and many other islands of a large archipelago, on the equator. The capital city is Jakarta, situated on Java.

Industrial Revolution

The name given to the age of industrial growth in eighteenth and nineteenth century Europe. It came about through the growth of scientific ability at manipulating the earth's resources, especially coal and iron.

Mt. Fuji, Japan. © *Paul Harris, 1993*

Billabong, Australian outback. © *Kieran O'Mahony, 1993*

Tundra topography, Russia Far East. © *Paul Harris, 1993*

Volcanoes - Pacific Ring of Fire, Kamchatka. © *Paul Harris, 1993*

Industrialist

In economic geography a person who makes a living in the pursuit of business at the tertiary level of events. The primary level includes farming, fishing, mining and forestry. The secondary level involves the processing and manufacturing of primary materials into a marketable finished product. Employment in the tertiary sector is concerned with the distribution and sale of finished products and the public services. This includes transport and communications, banking, insurance and administration.

Infrared waves

When the wavelength of light is smaller than that of visible light it can fall into a number of other categories, one of which is infrared light.

Inner city decay

Many older cities that grew up in the industrial revolution had older brick buildings in the inner city and these were abandoned with the development of the automobile and trains for suburbia. The inner city was thus abandoned, except for business. This was termed inner city decay. Today the trend is being reversed and developers are attracting people, especially families, into the inner city to inject a modicum of life back into the city.

Inselberg

A spectacular erosional feature (rock-island) called a ventifact since it was created by the abrasive power of the wind.

Insolation

The ability of the sun to heat up the land. Aspect has a direct impact on the amount of warmth a place receives. For instance if your house is situated on a slope that is facing towards the sun then it has a good chance of receiving insolation.

Interglacial periods

Long periods of time when temperatures rise, the ice-sheet retreats and a general melt takes place.

Interlocking spurs

In a youthful river valley, a view upstream will reveal the interlocking spurs. A spur is the name of the slope that extends from a hill into the lowlands.

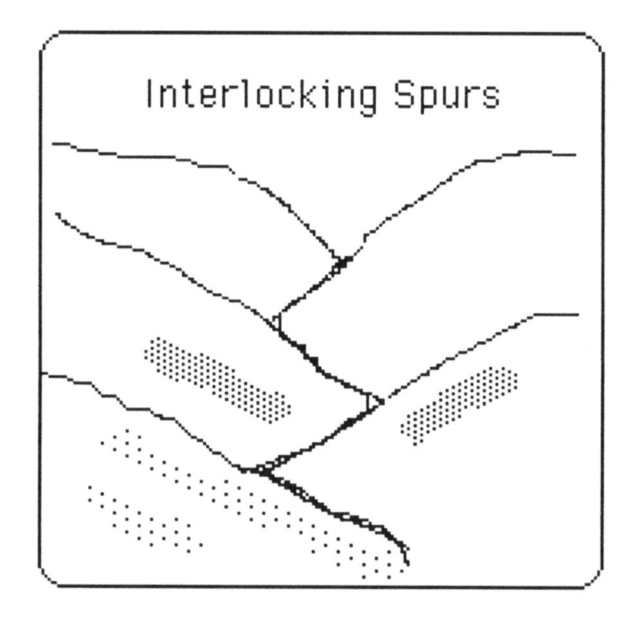

I.iii Interlocking Spurs

International co-operation

In today's global village international co-operation is the term that is used to explain the coming together of two or more opposing political nations for the common good. Nuclear disarming is a good example of international co-operation.

International date-line

An imaginary line that separates the new day from the previous day. The date to the west of the line is one day earlier than to the east. It follows the 180th meridian on the globe, which is opposite Greenwich Mean meridian. It conveniently lies between America and Asia since it passes through the Bering Sea and it extends on down into the South Pacific. This line digresses from the perpendicular in places to accommodate island groups like the Aleutian chain.

International Monetary Fund

An agency that assists member states achieve economic well-being and improved standards of living.

Intrusive igneous rock

A plutonic rock that cools down after a period of volcanic activity.

Ionosphere

A region of the upper atmosphere that contains many electrically charged particles called ions, because solar energy is absorbed by the gases in the atmosphere. The ionosphere is utilized for sending radio waves because it bounces these waves back down to the surface of the earth.

Iowa

A state in central U.S. Des Moines is the capital city. Also a river flowing into the Mississippi.

Iran

A country in southwest Asia, bordered on the north by the Caspian Sea and on the south by the Persian Gulf. Teheran is the capital city. With its history dating back 5,000 years it is said to be one of the world's oldest countries.

Iraq

A country in southwest Asia at the head of the Persian Gulf. Baghdad is the capital city. Iraq stands on the site of the world's first civilization, which developed in Mesopotamia - the land between the rivers Tigris and Euphrates.

Ireland

An island on the western edge of Europe close to Britain. Dublin is the capital city.

I.iv Ireland

Iron

A heavy metal that was one of the mainstays of the industrial revolution.

Iron Age

A period in history between 1500 and 1000 B.C. when the use of iron was widespread for making tools and implements.

Iron Pyrites

Also called "fool's gold" it is a mineral that closely resembles gold but is comparatively worthless.

Iroquois Indians

A powerful confederation of tribes that formed the most efficient Native American organization known as the *Iroquois Long House.*

Irrigation

A method of transporting water to agricultural land by artificial means. Thousands of acres that would normally be unproductive desert-scape are producing fruit and food for the urban markets because of irrigation technology.

Island

A landmass, smaller than a continent, surrounded by water.

Islet

A small island usually in a river channel.

Isobar

A line drawn on a weather chart joining together places that have the same atmospheric pressure. Iso means equal, in Greek, and bar means weight.

Isolationism

The philosophy that we do not need the rest of the world.

Isostasy

Upward and downward movement of the land relative to the sea level.

Isostatic recoil

The slow rising of land masses relative to the sea level after an ice age. This results in many features, including raised beaches, drowned river valleys called rias and drowned u-shaped valleys called fjords.

Isotherm

An imaginary line on a map that joins places with the same temperature.

Israel

A country bordering on the eastern Mediterranean, that makes up most of the Holy Land as referenced in the Bible. Jerusalem is the capital city. Israel was founded in 1948 as a homeland for Jews from anywhere in the world.

Istanbul

The largest city in Turkey on the Bosporus and the Sea of Marmara. It was formerly called Constantinople. Sited on the crossroads of two continents, Asia and Europe, Istanbul has been one of the world's most important cities throughout history.

Isthmus

A narrow neck of land that connects two larger areas of land. An example of an isthmus is Panama, since it connects North America and South America.

I.v Italy

Italy

A country shaped like a boot in southern Europe that projects into the Mediterranean between the Adriatic and the Tyrrhenian seas. It contains a number of islands including Sicily, Sardinia and other smaller ones. Rome is the capital city.

Jackson

The capital city of Mississippi.

Jamaica

An island country in the West Indies. The capital city is Kingston.

Japan

A country in east Asia comprising a number of islands, including Honshu, Hokkaido, Kyushu and others. They are situated in the western Pacific. The capital city is Tokyo.

Jerusalem

The capital city of Israel and a holy city for Christians, Jews and Muslims.

Jet Stream

A prevailing high altitude wind that circles the globe. Up here in the atmosphere, the geostrophic wind blows at speeds as fast as three hundred miles per hour, unimpeded by land masses or land configuration. A jet stream can speed up or slow down long aircraft journeys and is referred to as head-wind or tail-wind. See also; Geostrophic.

Jordan

An Arab kingdom in the Middle East. The capital city is Amman. The river Jordan divides the country into two halves - the West Bank and the East Bank.

Juneau

The capital city and a port in Alaska situated in the southeast along a narrow coastal strip.

Jungle

A dense tropical forest region.

Jupiter

A celestial body that is one of the nine planets of the Solar System. Jupiter is roughly eleven times as large as the earth.

Jutland

The peninsular portion of Denmark.

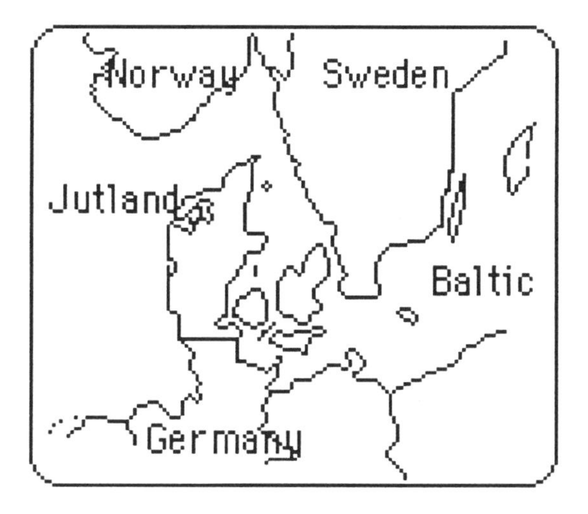

Fig J.i Jutland

Kabul

The capital city of Afghanistan on the river of the same name.

Kalahari

An arid desert in southwestern Africa inhabited sparsely by Bushmen. The desert is in the region corresponding to the zone of descending dry air from the tropics.

Fig. K i Kalahari

Kamchatka

A long narrow peninsula between the Bering Sea and the Sea of Okhotsk, in the Russian Far East. Part of the Pacific Ring of Fire, a range of volcanic peaks extends along the peninsula. Cold Arctic currents make for harsh living conditions. Hunting, fishing and reindeer herding are the main livelihoods for the indigenous peoples - Koryak, Chukchi, Eveni and Itel'men.

Kampuchea/Cambodia

A country in Southeast Asia bordering with Thailand, Laos and Vietnam. The capital city is Phnom Penh.

Kansas

A state in central U.S. midway between the Atlantic and the Pacific. Topeka is the capital city. Kansas consists of a flat tract of land in the south central United States, noted for its rich soils that were created by wind blown deposits called loess. Also a river that flows into the Missouri.

Karst

A name given to limestone topography, called after a barren region in

Yugoslavia. Typically, a karst region displays underground drainage, sink-holes and caverns.

Katmandu

The capital city of Nepal. It marks the focal point of many ascents on Mount Everest, the highest peak in the world.

Kenai

A peninsula in south Alaska east of Cook Inlet.

Fig. K ii Kenai

Kentucky

A state in eastern U.S. Frankfort is the capital city.

Kenya

A country of east Africa bordering on the Indian Ocean. Nairobi is the capital city.

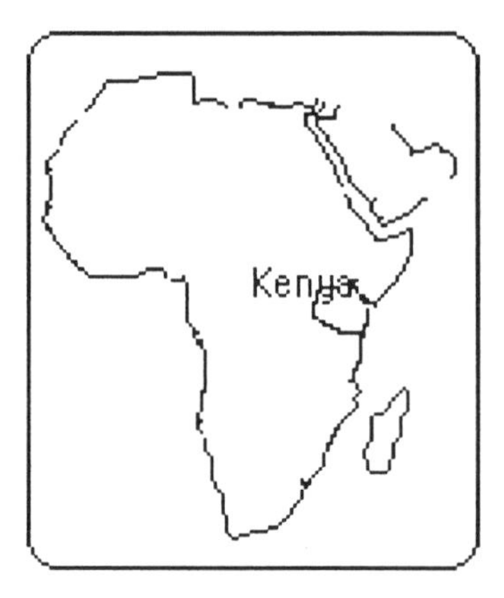

Fig. K iii Kenya

Kettle lake

A lake that is formed, often between drumlins, in terrain where glacial deposits have been left behind. Sometimes large blocks of ice remained as the ice retreated. When they melted they left bowl-shaped holes often occupied by lakes. Most of the shallow lakes of Michigan and Minnesota are kettle lakes. See also; Drumlins.

Kiel Canal

A canal in northern Germany, one of the busiest in the world, that links the Baltic with the North Sea. It is situated on the lower "neck" of Jutland inside the border with Germany.

Kilimanjaro

A volcanic cone that lies on the equator, on the border of Kenya and Tanzania, in East Africa. It is almost 20,000 feet in elevation and is snow covered even though it lies on the equator. It is the highest mountain in Africa.

Killimanjaro

Knickpoint

A definite mark in the long profile of a river depicting a change in gradient that could result from differential erosion of varied bands of geological strata.

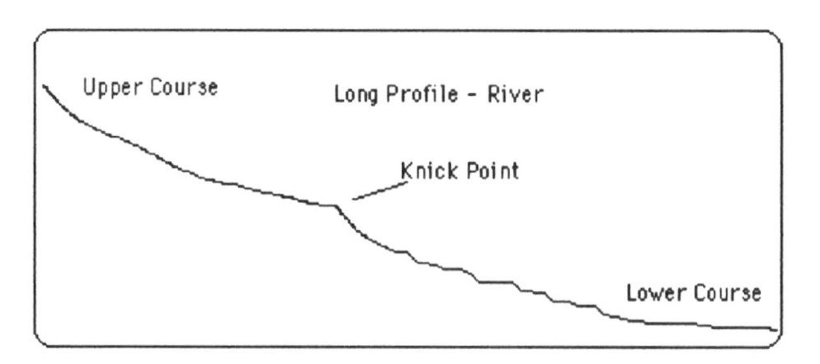

Fig. K iv Knickpoint

Korea

A country on the peninsula of the same name, divided after World War Two into North and South Korea. Seoul is the capital of South Korea. Pyongyang is the capital of North Korea.

Krakatoa

A volcano on an island of the same name, between Sumatra and Java. It erupted in 1883, and caused one of the world's worst disasters. The eruption triggered a tidal wave that was responsible for the deaths of 36,000 people on the neighboring island regions. It also caused vivid sunsets for many years since dust from the eruption penetrated the upper atmosphere.

Kuwait

An oil-rich country in southwest Asia in the Persian Gulf. The capital city is also called Kuwait.

Labrador

The large peninsula in northeast Canada, situated between the Atlantic and Hudson Bay.

Labrador current

The icy Arctic current flowing south from Baffin Bay past Labrador and Newfoundland into the Gulf Stream. The mixing of the warm and cold currents at this point creates a phenomenon known as the Grand Banks fishing grounds. This is an area of continental shelf where the plankton is churned up and shoals of fish forage for food.

Lacolith

An intrusion of igneous rock between the layers of sedimentary strata causes an upfold or bulge known as a lacolith.

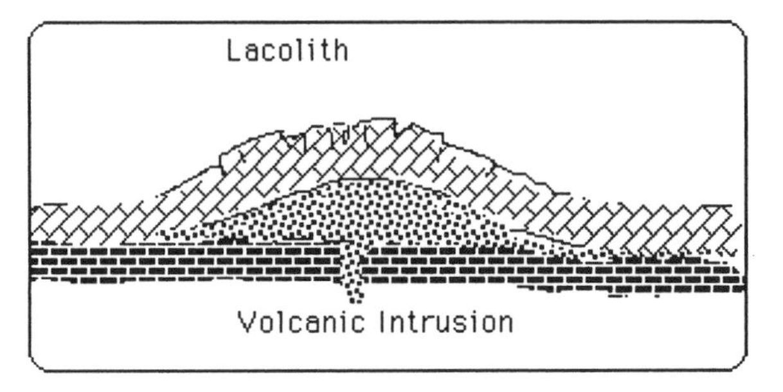

Fig. Li Lacolith

Lacustrine delta

A depositional feature formed when a river enters a lake and slows down, thereby depositing its load.

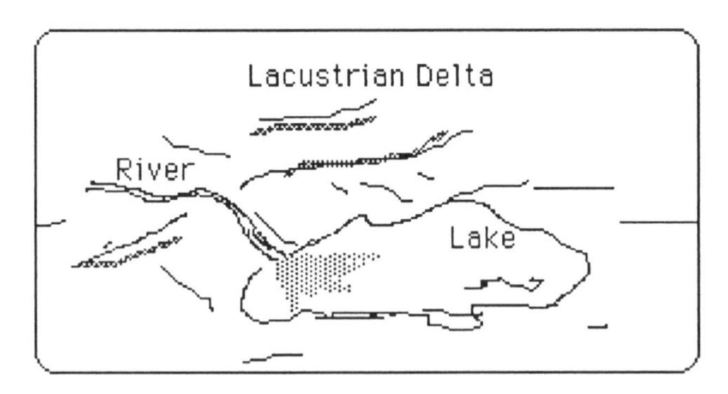

Fig. Lii Lacustrine delta

Lake

An inland body of usually fresh water, generally formed by some obstruction in the course of flowing water. Typically morainic dammed lakes are formed by glacial deposition, and rock basin lakes are formed by glacial erosion. Lakes have a geographic life process. Since rivers bring sediment and other deposits in their deltas the lake is filling up on one side, and since the outlet stream or river is eroding the lip on the other side it is only a matter of time before the lake disappears. Some lakes have no visible outlets like rivers or streams, but water is lost through evaporation and seepage.

Lacustrine delta

Land breeze

An air mass blowing seaward from the land, usually occurring in the late afternoon when the heat of the sun is lessened and the land begins to lose its heat faster than the water. Because the air over the sea is still rising, the cooler air from the land blows seaward to take its place.

Land Fill

The disposal of waste material and garbage by burying it under a shallow layer of soil.

Land Sat

A satellite that sends weather and meteorological information back to earth from its orbit high above the planet.

Land slide

The slipping of a mass of loosened rocks and earth down a steep hillside.

Laos

A country located in Southeast Asia. The capital city is Vientiane.

Lapland

The region of northern Europe inside the Arctic Circle, inhabited by the Lapps, an Eskimo-like people that live in the northern tundra.

La Paz

Bolivia's largest city is the highest city in the world, at 12,000 feet.

Las Vegas

A tourist attraction and city in Nevada.

Lateral erosion

This occurs in river channels when the sides of the beds are abraded by hydraulic action.

Lateral moraines

These are the lines of deposition that lie on the surface of the glacier as it advances down the valley, constantly receiving debris, rock fall and detritus from the steep sides of the valleys about it.

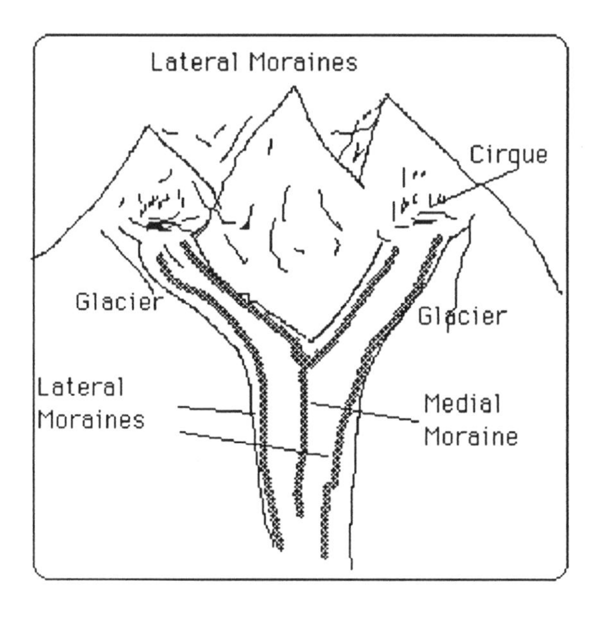

Fig. Liii Lateral Moraines

Latin America

The Spanish and Portuguese speaking countries of North and South America. Although Mexico and Central America are geographically linked with North America, they are culturally related to Latin America.

Latitude

Angular distance north or south of the equator. It is a means of locating places.

Latitude, line of

An imaginary line drawn on a map parallel to the equator that joins places that are equidistant from the equator. Angular distance is measured in degrees north and south of the equator.

Latvia

A country in north central Europe bordering on the Baltic Sea. Riga is the capital city.

Lava

Molten hot rock pouring out of a volcanic vent.

Law of Gravitation

The mathematical formula that identifies the bonds between the celestial bodies.

Laws of planetary motion

Three laws that govern all moving bodies.

Leaching

The process by which rain water filters minerals through the soil into a hard pan in the horizon below. The addition of fertilizer to the soil and a process known as deep ploughing compensates for leaching.

Lebanon

A country of southwestern Asia bordering on the Mediterranean Sea. Beirut is the capital city.

Legend

The explanation on a map for the symbols, colors and scales used to represent the real world.

Lesotho

A poor and mountainous country in the middle of South Africa. The capital city is Maseru.

Levee

An embankment, either natural or man-made, built alongside a channel to prevent high water flooding onto the surrounding land.

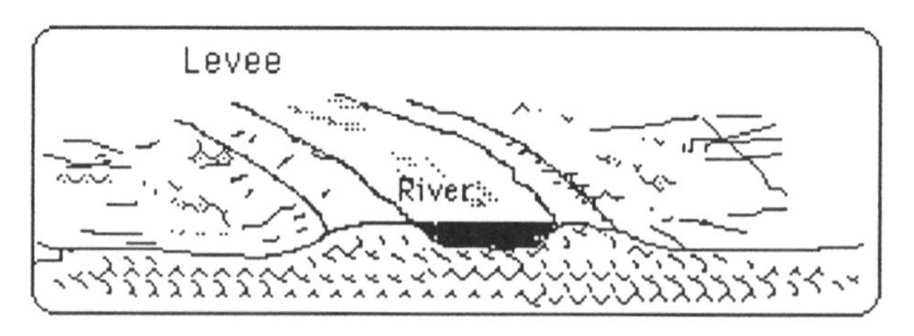

Fig. Liv Levee

Lewis & Clark

In 1804, these two explorers left St. Louis and made their way up the Missouri River, crossed the Rockies and followed the Columbia river to the Pacific coast.

Liberia

A country in West Africa. It is the oldest independent black nation in Africa. Monrovia is the capital city. An American charitable society founded Liberia in 1822 to provide a safe haven for freed black slaves.

Libya

A country in northern Africa bordering on the Mediterranean. Tripoli is the capital city.

Lichen

A moss that grows on rocks and trees and is a good indication of clean air.

Liechtenstein

One of the smallest countries in the world located in south central Europe between Austria and Switzerland. The capital city is Vaduz.

Light year

The distance light travels in one year at 86,000 miles per second.

Lightning

Charged particles in the ionosphere are attracted and in certain conditions, especially violent storms, they will plunge to earth in a bright flash.

Limestone

A light colored sedimentary rock that is laid down in the sea bed over millions of years and is composed of decayed marine plants and animals.

Lisbon

The capital city and port of Portugal, situated on the estuary of the Tagus river.

Lithosphere

The top layer of the earth that has cooled from molten magma. Lithos in Latin means rock.

Lithuania

A country on the eastern side of the Baltic Sea. The capital city is Vilnius.

Load

The volume of detritus silt and sand that is carried in suspension and by saltation by the river.

Loam

A mixture of sand and clay to make a rich agricultural soil.

Location

The site or situation of a town, farm, or industry, etc.

Loch Ness

A lake in Scotland, believed to contain a dinosaur-like reptile known as the Loch Ness Monster.

Loess

Wind blown silt that is deposited over a number of centuries to form a rich agricultural soil.

London

Capital city and cultural center of England situated on the river Thames.

Long-shore drift

The movement of silt and sediment along the shore caused by the configuration of the coast line and the direction of the incoming waves. Sometimes this will result in a build up of material on one side of a beach or lead to the formation of a sand spit.

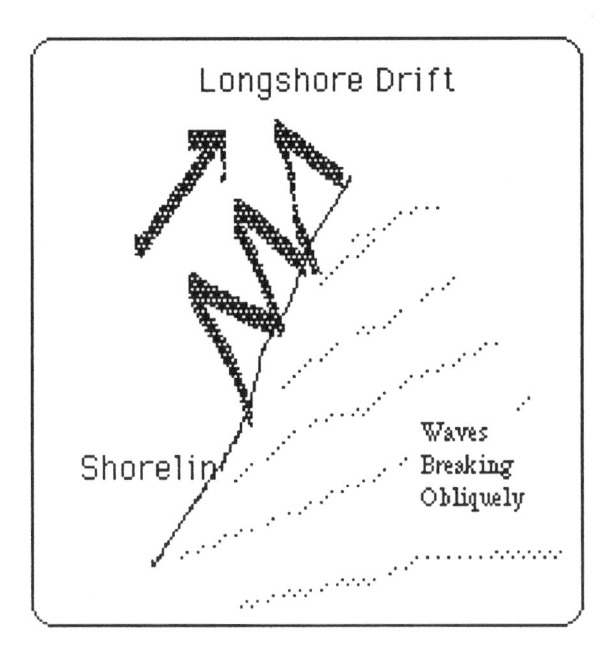

Fig. Lv Long-shore Drift

Longitude

Angular distance east or west of the Prime Meridian. A line of longitude is an imaginary line on a globe, measured in degrees east or west of the central meridian at Greenwich. Together with lines of latitude, it is a means of defining the location of places.

Los Angeles

A city and port in southern California. The third largest city in the U.S. It is a vast sprawling urban conglomeration of ethnic diversity and local communities.

Louisiana

A state in southern U.S. Its capital city is Baton Rouge.

Low pressure

Hot air rising causes low atmospheric pressure and is usually associated with rain and inclement weather.

Lower course

This is the final course of a typical river pattern. It is characterized by sluggish movement, deposition of sand and silt, and distributaries that lead through large deltas.

Lunar eclipse

The obliteration of the moon by the earth's shadow.

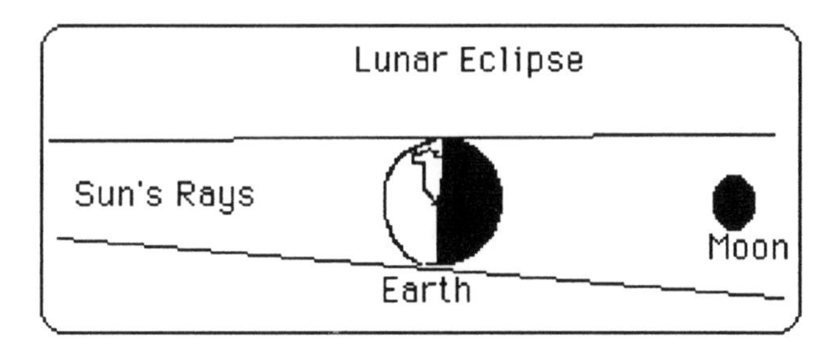

Fig. Liv Lunar Eclipse

Luxembourg

One of the smallest countries in western Europe, located between Belgium and Germany. Luxembourg is the capital city.

Machu Picchu

A region in southeastern Peru on a mountain side near Cuzco, once inhabited by the Inca people.

Mackenzie River

Canada's longest river. One of its many tributaries joins it from Alberta and eventually the Mackenzie spills into the Beaufort sea after 2,500 miles in the Arctic north. The Mackenzie receives water from many large lakes including Athabasca, Great Slave and Great Bear.

Madagascar

A large island country in the Indian ocean off the coast of Malawi in east Africa. The capital city is Antananarivo.

Madrid

The capital city of Spain, situated in the heart of the country.

Magellan

The man who lead the expedition that was to circumnavigate the globe, in 1521. This was the first definitive proof that the earth was not flat.

Magellan, Strait of

The narrow strait at the tip of South America separating Tierra del Fuego from the mainland.

Magma

Hot molten rock that builds up in chambers inside the lithosphere - below the crust.

Magma chamber

The accumulation chamber of hot molten magma in the earth's crust.

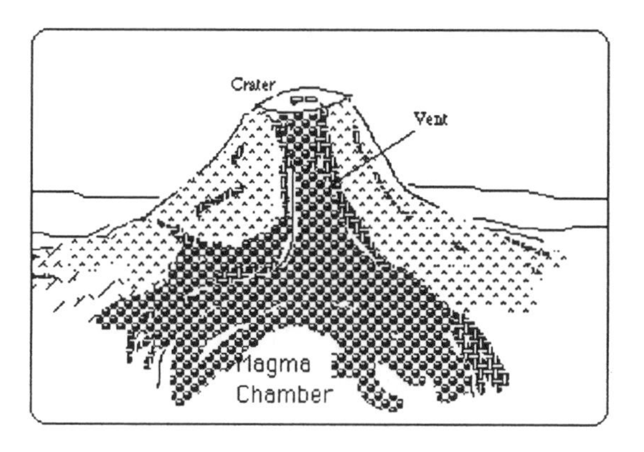

Fig. M i Magma Chamber

Magnetic field

The force field that surrounds our planet earth.

- pole

The north or south pole of our magnetic field.

Maine

A state in northeastern U.S. in New England. Its capital city is Augusta.

Malawi

A long narrow country on the western shores of Lake Nyasa in southeastern Africa. The capital city is Lilongwe.

Malayan Peninsula

A peninsula situated in southeastern Asia.

Malaysia

A tropical country in Southeast Asia, partly located on the Malayan peninsula and the remainder on the northern part of the island of Borneo. It is one of the chief natural rubber producing regions in the world. The capital city is Kuala Lumpur.

Maldives

One of the smallest independent countries in the world, consisting of approximately 1200 small coral islands in the Indian Ocean. Because of the strategic location, fishing and tourism are the main sources of employment for the island inhabitants. The capital city is Male.

Mali

A sparsely populated semi-desert country in western Africa. The capital city is Bamako and the city Timbuktu is also in Mali.

Malta

A densely populated island country (formerly, a British colony) in the Mediterranean Sea. The capital city is Valetta.

Malthus

A nineteenth century English economist and geographer who predicted that the earth could not sustain its population. He maintained that food supplies would not grow at the same rate that population increased.

Malthusian Checks

These were certain predicted outcomes of the fact that world food supplies would not be sufficient to sustain the growing population numbers. Malthus said that famines, diseases and war would be the agents by which the earth could control population growth naturally.

Manchuria

An ancient region of northeastern China.

Manhattan

An island and a borough in New York City. It is renowned as a commercial and cultural center of the world. It boasts many famous landmarks, theaters and tourist attractions, including Times Square, Broadway and the Empire State Building.

Manila

The capital and largest city of the Philippines.

Manitoba

A Canadian province in the prairie belt. The capital city is Winnipeg.

Mantle

The top layer of the lithosphere (just below the earth's surface), usually between eight and twenty miles in thickness.

Maoris

Polynesian descendants that are the native inhabitants of New Zealand.

Map

A graphic representation of the real earth (or a part thereof) on paper, drawn to scale.

- general reference map

Ordinary road or street map.

- thematic map

A map that represents a particular theme, for instance the rainfall in a certain region.

- projection

One of several methods of representing a sphere on a flat surface like a sheet of paper. Typically, a transparent globe is used with a light inside that projects the lines of latitude and longitude onto a paper surface.

- scale

The relation between distances on the earth and on the map.

Marble

A strong and useful building rock. It is a derivative of limestone that has been metamorphosed into another form due to heat or pressure caused by folding or volcanic activity.

Mariana Trench

The deepest place on the earth's surface. It is a deep valley in the ocean floor, situated off the Marianas in the south Asian sea.

Maritime climate

A climate that has a very noticeable maritime influence on it. The sea has a modifying effect on the landmasses that are adjacent to it.

Marmara, Sea of

A body of water in Turkey that is connected to the Black Sea by the Bosporus and to the Aegean Sea by the Dardanelles.

Mars

The next planet to earth and the fourth closest to the sun, in our solar system. One Martian year is nearly twice that of earth, since it takes nearly 690 days to orbit the sun. Speculation about life on Mars exists because surface conditions are so similar to those on earth.

Marseilles

City and port in the south of France on the estuary of the river Rhone.

Maryland

A state in eastern U.S. The capital city is Annapolis.

Massachusetts

A state in eastern U.S. The capital city is Boston.

Matterhorn

A pyramidal peak in the European Alps between Switzerland and Italy. It is a well known tourist attraction in Switzerland because of its spectacular beauty and a coveted challenge to climbers who usually make ascent plans from the Hornli hut on the glacier. See also; Mountains, Alps.

Maui

One of the islands of the Hawaiian chain in the Pacific Ocean.

Mauna Loa

A volcano on Hawaii island.

Meander

Meander

Rivers develop meanders usually in their middle and lower courses. They rarely experience a smooth laminar flow and as a result friction with the sides and bed causes turbulences that result in differential erosion. Patterns are established and undercutting takes place on one side while slip-off slopes develop on the other side so that the river begins to migrate across the flood plain. Large meander loops take the river on circuitous journeys through the valley. At times the loops are cut off to form stagnant lakes called ox-bow lakes.

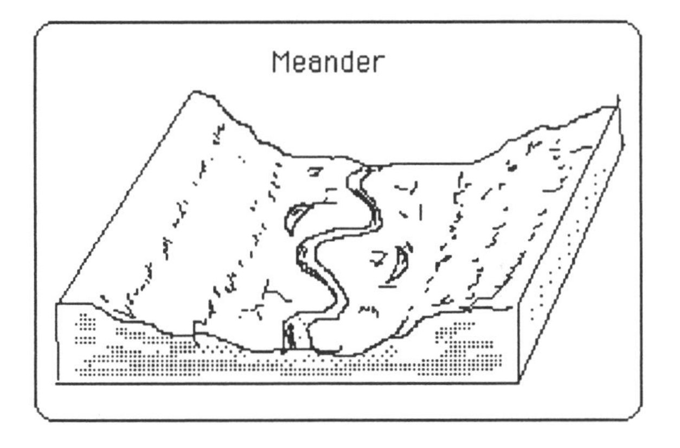

Fig. M ii Meander

Mecca

A holy city, located in western Saudi Arabia, and a place of worship for Muslims who practice the religion of Islam.

Medial moraine

When two lateral moraines coalesce from tributary valleys, a medial moraine is formed. See also; Lateral Moraine.

Mediterranean Sea

The body of water that separates southern Europe from north Africa. It is the largest inland sea in the world and an important trade link between the Middle East and Europe.

Melbourne

The capital city of Victoria in South Australia.

Melting - ice

Glaciers advance until the climatic temperature rises above freezing at which point they melt.

Meltwater

This is a voluminous quantity of water that is created when glaciers and ice sheets melt and change back into run-off.

Mercator

A visionary geographer who created map projections that were used over many centuries for navigation and teaching geography.

Mercury

A heavy metallic element.

- barometer

The instrument used to measure the pressure of the atmosphere. A normal atmospheric pressure would sustain a column of mercury in a tube to a height of about 30 inches. Falling barometric pressure signifies rain and stormy conditions. Rising pressure signifies fair stable weather.

Meridian

An imaginary line on a map that is measured in degrees east or west of the central line at Greenwich. See also; Longitude.

Mesopotamia

A region in southwest Asia between the Tigris and Euphrates rivers where early man first practiced agriculture. The name Mesopotamia comes from the Greek word which means *between the rivers.*

Mesozoic era

One of the prime time eras in geologic history associated with the emergence of plants and animals on the earth.

Metamorphic rock

Rock that has undergone intense change due to heat and pressure from either tectonic movement or vulcanism. For example, marble is formed from limestone.

Meteor

A celestial fragment that hurtles through space often colliding with other planets. They appear as shooting stars in our atmosphere as they are subjected to intense heat.

Meteorite

A meteor that falls to earth is called a meteorite.

Meteorology

The study of earth's atmospheric conditions that cause weather.

Mexico

A country south of the U.S. border. The capital city is Mexico City.

Miami

A city and port in southeast Florida situated on Biscayne Bay.

Michigan

One of the Great Lakes of North America. Also a state in the same region whose capital city is Lansing. Michigan borders four of the Great Lakes.

Middle course

The section of a river profile when the valley widens out, the gradient eases up, and meanders begin to form.

Middle East

The name given to a large geographical region extending from northeastern Africa to western Asia. Among the countries included are Egypt, Kuwait, United Arab Emirates, Israel, Iran, Iraq and Turkey.

Mid oceanic ridge

A tectonic fault where new rock is constantly emerging, as the ocean floor is spread apart. Island chains of volcanic origin, like Surtsey and the Azores, dot this mid oceanic ridge.

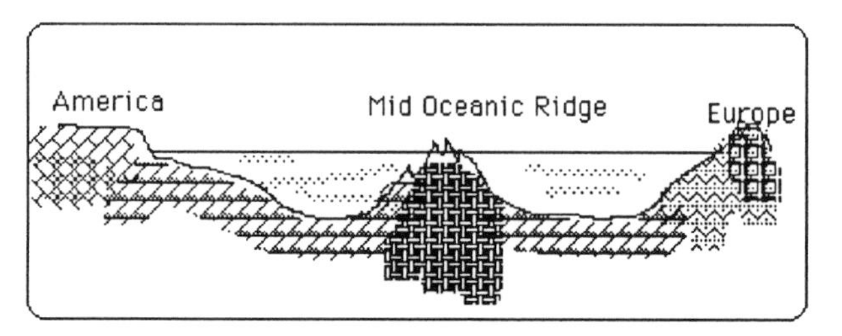

Fig. M iii Mid Oceanic Ridge

Midnight sun

A phenomenon of the Polar regions where the sun never sinks below the horizon on and around midsummer's day.

Midwest

A place in the U.S. between the west and the midlands, that is distinguished by prairie lands and mountains.

Milky Way

The galaxy that contains the solar system where the earth resides. It appears as a broad faintly luminous arch across the night sky that contains many billions of stars and clouds of interstellar gas.

Mineral

An element in a rock.

Minnesota

A state in the Midwest. The capital city is St. Paul.

Mississippi

The river that rises in Lake Itasca, in Minnesota, flows to the sea in the Gulf of Mexico and drains the Midwestern states of North America. Also a state in southern U.S. whose capital is Jackson.

Missouri

A river flowing from southwestern Montana to the confluence with the Mississippi. Also a state in central U.S. whose capital is Jefferson City.

Moho

The line that divides the earth's crust from its mantle, called after the Yugoslavian geologist. See also; Mantle.

Mojave

A desert region in southern California. It is used as a landing strip for space shuttles and other craft that require long runways and flat surfaces.

Mombasa

A large port and city in Kenya, East Africa.

Mongolia

A country in eastern Asia, north of China. It comprises a large tract of sparsely peopled land, including the Gobi desert, settled by nomadic tribesmen. The capital city is Ulan Bator.

Monsoon

A strong wind that is caused by differential heating of land and sea and is associated with seasonal heavy rain conditions in the Indian Ocean environs.

Montana

A picturesque state in northwestern U.S. The capital city is Helena.

Montreal

A major center of Canadian commerce and a leading inland seaport. This French speaking city lies on an island at the confluence of the Ottawa and St. Lawrence rivers, in southern Quebec. It boasts the largest underground shopping complex in the world.

Moon

A large celestial body that orbits around the earth approximately once every 28 days.

Moraine

A glacial feature that results in the deposition of boulder clay, morainic dammed lakes and outwash sands and gravel.

Morocco

A country in northwestern Africa on the Atlantic. The capital city is Rabat.

Moscow

The capital city of Russia situated on a river of the same name.

Mount Everest

The highest mountain in the world. See also; Everest.

Mount Kilimanjaro

A volcanic peak on the equator in East Africa, bordering Kenya and Tanzania. See also; Kilimanjaro.

Mount McKinley

The highest mountain in North America. The mountain is part of the Alaskan Range and rises to a height of 20,000 feet.

Mount Rainier

A volcanic peak in Washington State.

Mt. St. Helens

A volcanic peak in Washington State, that erupted in 1980 with great destructive force.

Mountain formation

Mountains are formed in a number of ways. Tectonic movement causes folding when land masses are crushed against each other. Subduction zones cause eruptions and volcanic peaks to be thrown up. Other structural movements and pressure result in block mountains being raised up. See also; Subduction Zone.

- Ranges

- Alps

A mountain block in central Europe particularly in northern Italy and Switzerland.

Pyramidal Peak, Arete, Alpine topography

- *Andes*

A mountain range that forms the backbone to South America stretching from Peru to Tierra del Fuego.

- *Appalachians*

Mountain ridge and valley topography in the eastern United States. It is breached by a number of corridors through which routeways have been established.

- *Himalayas*

A mountain range in south central Asia containing the highest peak in the world - Mt. Everest in Nepal.

Mouth - river

Where the river enters the sea, usually through an estuary but sometimes through distributaries in a delta or over a bar.

Mozambique

A country in southeastern Africa, on the Indian Ocean. The capital city is Maputo.

Nadir

That point opposite the zenith and directly below the observer.

Nagasaki

A seaport in western Japan that was virtually destroyed at the end of the second world war by a nuclear bomb.

Nairobi

The capital city of Kenya in East Africa. See also; East Africa, Kenya.

NASA

An acronym for National Aeronautics and Space Administration, the agency for the exploration of the universe.

Nashville

The capital city of Tennessee and the center for the country music industry.

National grid

The grid that cartographers use to denote maps for specific parts of the nation.

National Park

An area of scenic beauty, historical or scientific interest maintained and preserved by the Federal Government for the public to visit.

Native American

The name referring to the Indian tribes-people who are endemic to the United States of America.

Natural corridors

Route-ways and channels of movement produced by nature and later utilized by man.

Natural gas

A hydrocarbon fuel occurring naturally in the earth particularly in association with petroleum.

Natural resource

A form of wealth produced by nature that man can utilize for prosperity, for example, timber, oil.

- conservation

A policy of utilizing the natural resources that are available to man on the planet in such a way that they will not disappear altogether.

- nonrenewable

Some natural resources will disappear when they are used up. Coal, peat, oil and other hydrocarbons are examples of nonrenewable resources.

- renewable

Some natural resources are renewable like timber and water, but care has to be taken not to exploit them to obliteration.

Navajo

The largest Indian tribe in the U.S.

Navigable

A body of water that is wide and deep enough to allow vessels travel on it without fear of going aground.

Neanderthal

The name given to a primitive human being based on the discovery of skeletal remains, dating back to the Paleolithic world, in a valley - the Neander Gorge - near Dusseldorf, in Germany.

Neap tide

This refers to the tide that occurs just after the first and third quarters of the lunar month, when the difference between high and low tides is the smallest.

Nebraska

A rich agricultural state in the Mid-western U.S. The capital city is Lincoln.

Nebula

Vast luminous cloud-like patches seen in the night sky, consisting of groups of stars or galaxies too far away to be seen clearly.

Neo-lithic

A period during the stone age when man discovered tools and developed agriculture.

Nepal

A country in the Himalayas between India and Tibet, underneath Mount Everest. Katmandu is the capital city.

Neptune

The eighth planet of the solar system.

Netherlands, The

A small country in western Europe on the English Channel and the North Sea. It gets its name from the fact that most of the land is below sea level. As a result sturdy dikes are continuously maintained to keep the sea at bay. The capital city is Amsterdam.

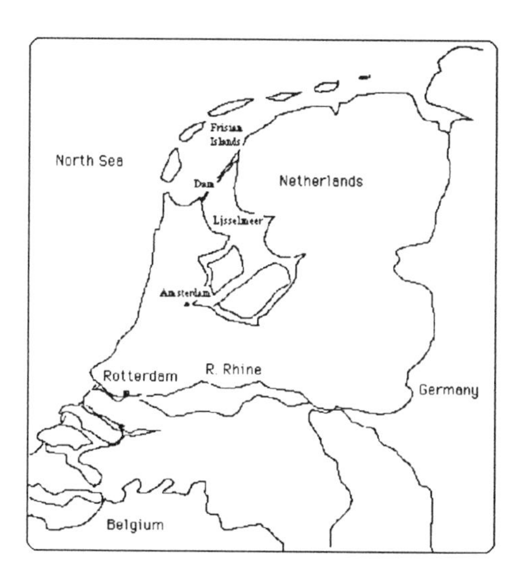

Fig. N i Netherlands

Nevada

A state in the Rocky mountains in western U.S. The capital city is Carson City.

Neve

The snow that accumulates on a glacier and is compacted on top of older snow from a previous season. Also known as firn.

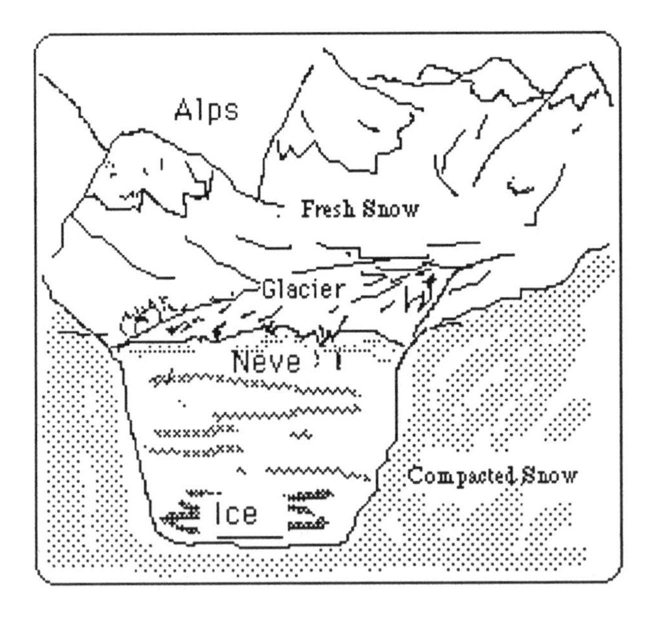

Fig. N ii Neve

New England

A geographic region in the north-eastern U.S. comprising the states of Maine, New Hampshire, Vermont, Massachusetts, Rhode Island and Connecticut.

New Hampshire

One of the thirteen original states of the U.S. located in New England. The capital city is Concord.

New Jersey

An industrial state in the northeastern United States. The capital city is Trenton.

New Mexico

A farming state in the southwestern United States. The capital city is Santa Fe.

New moon

A phase in the moon when it is between the earth and the sun and the dark side of its disk faces the earth. After two or three days it reappears as a thin crescent curving toward the right.

New Orleans

A city and port in southeast Louisiana between lake Pontchartrain and the Mississippi. It is famous for its music and food which is influenced by the Creole culture.

New South Wales

A state in southeastern Australia bordering on the Pacific. Sydney is the capital city. See also; Australia.

New York

A state in eastern U.S. The capital is Albany. Also a city and port on the Hudson river, that includes boroughs of the Bronx, Queens, Manhattan, Brooklyn and Richmond.

New Zealand

A country made up of two large islands southeast of Australia, in the south Pacific. The capital city is Wellington.

Newfoundland

A province in eastern Canada comprising the island of Newfoundland and Labrador. The capital is St. Johns.

Niagara Falls

A picturesque waterfall on the Niagara river in upstate New York that flows from Lake Erie north to Lake Ontario. It is divided by Goat Island into the Horseshoe Falls on the Canadian side and the American Falls on the southern side.

Nicaragua

The largest country in Central America located between the Caribbean and the Pacific. Its capital is Managua.

Nigeria

The most populated country in Africa. It is located in central West Africa on the Atlantic coast. The river Niger empties into the Gulf of Guinea here through a widespread delta. The capital city is Lagos.

Night

The time when the earth is facing away from the sun, between sunset and sunrise.

Nile

A major river in Africa, and the longest in the world, that rises in the 'Mountains of the Moon' and flows through Lake Victoria and northward for more than four thousand miles into the Mediterranean Sea, in Egypt. For much of its length it creates a fertile valley bordered by deserts on the east and west.

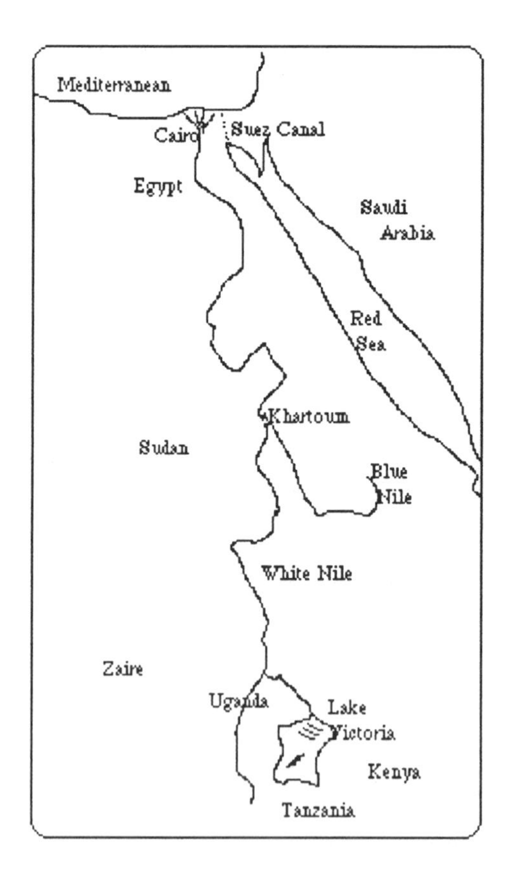

Fig. N iii Nile

Nomad

A pastoral dweller in barren and desert regions of the world, who moves from place to place as a means of subsistence. Nomad comes from the Greek - meaning a wanderer.

North America

Continent of the western hemisphere northwest of South America and bordered by the Atlantic, the Pacific and the Arctic Oceans.

North American Plate

Refers to the tectonic plate whereupon the American continent sits.

North Atlantic Drift

The warm ocean current that originates at the north equator and washes the shores of the Gulf of Mexico before turning north into the Atlantic, where it is called the North Atlantic Drift. See also; Gulf Stream.

North Equatorial current

The warm current that originates in the tropics and moves north into colder bodies of water.

North Carolina

One of the thirteen original states of the U.S. located in the Blue Ridge mountains. The capital city is Raleigh.

North Dakota

A farming state in the Midwestern U.S. The capital city is Bismarck.

North Germanic Plain

A post glacial plain of boulder clay that occupies the north of Germany.

North Pole

The northernmost point on the earth.

North Star

The star (Polaris) that is almost directly above the north pole. It can easily be located by sighting along the pointers of the Big Dipper.

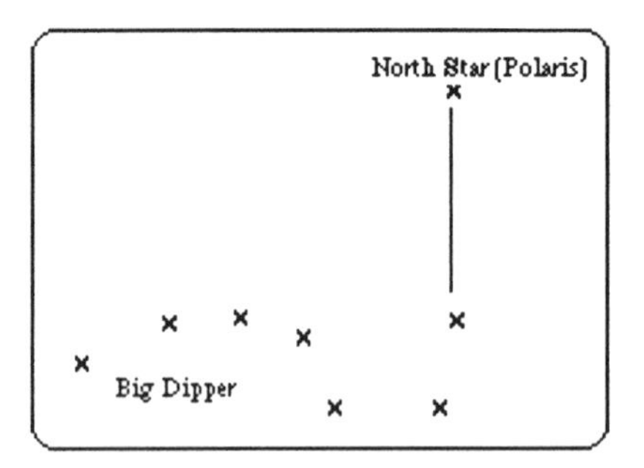

Fig. N iv North Star

Northern Hemisphere

The half of the earth that is north of the equator.

Northern Ireland

A region of northeast Ireland that is politically part of Britain. Belfast is the capital. See also; Ireland.

Northern Territory

A territory in central and northern Australia. Darwin is its capital. See also; Australia.

Norway

A country in northern Europe extending from the Baltic Sea to inside the Arctic circle. It is characterized by spectacular coastal fjords and steep mountains caused by isostatic recoil. Oslo is the capital.

Northwest Territories

A vast remote region in northern Canada. The capital city is Yellowknife.

Notch

An indentation in a cliff face caused by the onslaught of the sea, especially during storms.

Nova Scotia

A province in eastern Canada on the Atlantic coast. Nova Scotia comes from the Latin for New Scotland. The capital city is Halifax.

Nuclear waste

The disposal of radio active material that results from the generation of power from nuclear sources.

Nullarbor

A treeless desert bordering on the Great Australian Bight in southern Australia stretching from Adelaide to Perth. See also; Australia.

Nunataks

A mountain peak that protrudes through a continental ice sheet in the polar regions.

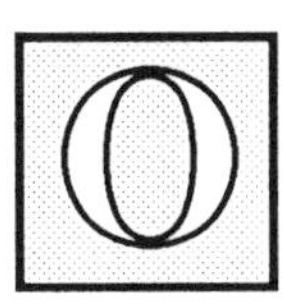

Oahu

An island of the Hawaiian chain in the Pacific, upon which Honolulu, Pearl Harbor and Waikiki are located.

Oak

A tree of the beech family that produces acorns.

Oasis

A fertile area in a desert region, usually around a water-hole, a well or spring, where palm trees and grasses can survive.

Oat

A cereal grass that is widely cultivated and used for producing food.

Ob

A river in western Russia that flows roughly 2500 miles north into the Arctic ocean.

Observatory

A building equipped for observation of natural phenomena especially astronomical observations.

Occident

Another name for the west, as opposed to Orient, the east. It is used when referring to the people who were born in Europe as opposed to Asia.

Occlude

When a cold front forces a warm front to lose contact with the surface of the earth it is said to be occluded and a wintry weather outlook is forecast.

Ocean

The body of salt water that covers nearly three fourths of the surface of the earth, often called sea. The three largest oceans of the world, the Pacific, the Atlantic and the Indian are part of one great ocean.

Ocean current

A warm or cold mass of water that is blown by the prevailing winds into other waters and modifies the climates of the countries that they encounter. Ocean currents are like great 'rivers' that move about in the oceans.

Ocean-floor-spreading

Tectonic movement and vulcanicity cause the upwelling of lava and new rocks in the mid oceanic ridges and the resulting ocean floor spreading. See also; Mid Oceanic Ridge.

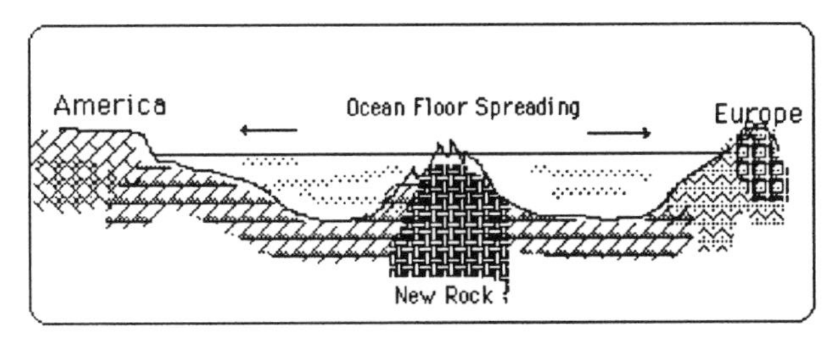

Fig. O i Ocean-floor-spreading

Oceanography

The science that deals with the oceans and includes the delimitation of their extent and depth, the physics and chemistry of their waters, marine biology and the exploitation of their resources.

Off-shore

Moving away from the shore toward the water. For example an off-shore wind blows out to sea.

Ohio

A state in the mid U.S. Columbus is the capital city. The Ohio river is a tributary to the Mississippi, and forms the southern border of Ohio State.

Oil

One of the resources for fuel that can be mined from the ground, on land or under the sea, and refined to burn in different combustible formats. It is also used in the manufacture of products.

Oil Rig, North Sea

Oklahoma

A state in the mid western U.S. The capital city is Oklahoma City.

Old Faithful

A geyser that erupts with regularity, every seventy-three minutes or so, in Yellowstone National Park.

Oman

A small country in the southeastern Arabian peninsula on the Arabian Sea.

It is one of the hottest places in the world. The capital city is Muscat.

Ontario

A province in Canada, located north of the Great Lakes. The capital city is Toronto. Canada's capital, Ottawa, is also located in Ontario.

On-shore

Moving from the water toward the shore. For example an on-shore breeze blows over the land near the shoreline.

OPEC

The Organization of Petroleum Exporting Countries. A conglomerate of oil producing countries that control the price and market for oil.

Orbit

The path described by one body in its revolution about another, as in the orbit of the earth about the sun.

Oregon

A state in the western U.S. situated on the Pacific coast. It is a leader in the lumber industry. Crater Lake, the deepest lake in the U.S. is located in Oregon. The Oregon coast is picturesque and rugged. The capital city is Salem.

Ocean vista, Oregon coast

Orient

The east. The inhabitants of Asia are described as being oriental. The word comes from the Latin for rising sun. See also; Occident.

Orography

That part of geography that deals with mountains.

Oslo

The capital city of Norway, lies at the head of the Oslo Fjord, where the Skagerrak and the Kattegat meet.

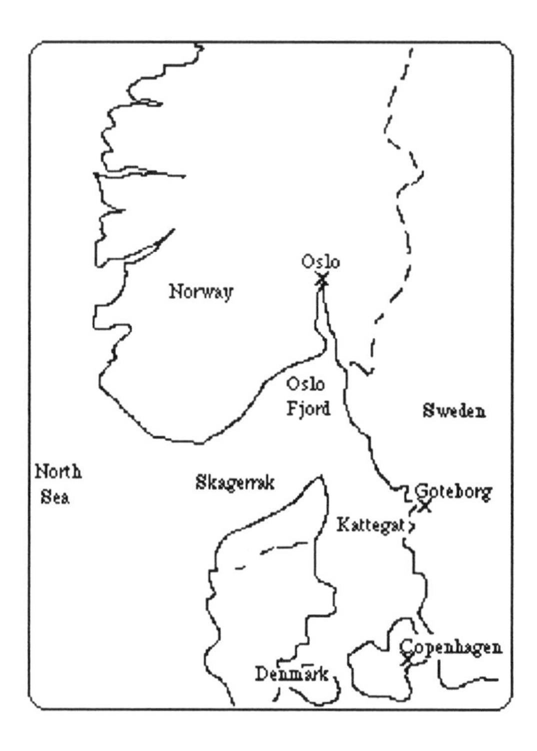

Fig. O ii Oslo

Orthographic projection

A map that is drawn with a particular technique showing a single view of the globe that is perpendicular to the view and the lines of projection.

Ottawa

The capital of Canada in southeastern Ontario.

Outlet

A stream flowing out of a lake or pond.

Outwash sand and gravel

The sediment and detritus deposited by rivers in the front of a melting glacier.

Overcast

When the sky is occluded by cloud cover it is said to be overcast.

Ox-bow lakes

The cut-off lakes when meanders abandon former courses and re-route in another channel. See also; Meander.

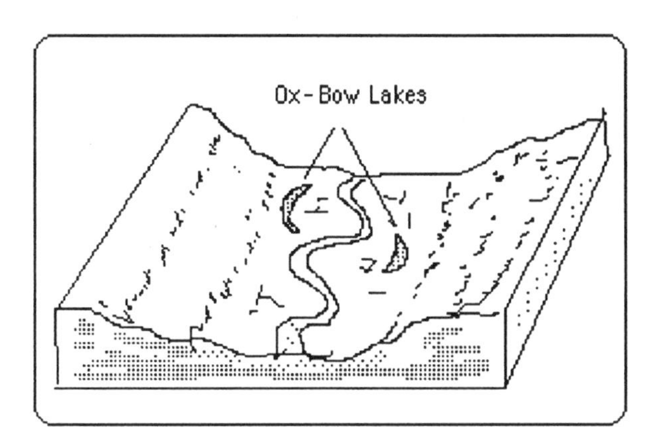

Fig. O iii Ox-bow Lakes

Oxford

An ancient university town on the Thames, England.

Oxygen

A colorless, odorless gas that constitutes about 21% of the atmosphere.

Ozone layer depletion

The destruction of the ozone layer that protects the earth from harmful solar ultraviolet light.

Pacific Northwest

A region in the northwestern U.S. It includes Washington, Oregon, Idaho and western Montana.

Pacific Ocean

The largest and deepest of the five oceans, covering approximately 30 per cent of the earth's surface. It is an extensive body of water that separates Asia from America.

Pacific plate

The tectonic plate upon which the Pacific ocean resides.

Pacific Ring of Fire

A name given to the phenomenon whereby volcanic activity occurs in a ring around the Pacific ocean.

Pakistan

A country of southern Asia bordering on India. The capital city is Islamabad.

Palaeozoic era

A geologic time when fishes and plants developed about one half million years ago.

Palestine

A region bordering the Mediterranean and extending to the river Jordan divided between Israel, Jordan and Egypt. It is the Holy Land to millions of Christians and Jews. Muslims also consider it a sacred place. It is the location of many of the events described in the Bible.

Panama

A country in south Central America. The capital city is Panama City.

Panama, Canal

A canal in south Central America that connects the Atlantic and Pacific oceans.

Papua New Guinea

A tropical island nation in the Pacific, north of Australia, that includes the Island of New Guinea and a chain of smaller islands. It is a member of the Commonwealth of Nations. The capital city is Port Moresby.

Paraguay

A landlocked country in central South America. Its capital is Ascuncion. Also a river that rises in the Mato Grosso in Brazil and flows south to the confluence with the Parana in Paraguay. Many dams harness electricity on the river Paraguay and generate income for communities and inhabitants throughout the country.

Paris

The capital city of France, and one of the foremost centers of culture in Europe. It is a major tourist attraction and visitors come from all over the world.

Pass

A natural gap, a wind gap, coll or break in a mountain ridge that provides a way through.

Patagonia

A barren region in southern Argentina and southern Chile between the Andes and the Atlantic.

Pater Noster lakes

Glacially scoured or morainic dammed lakes that resemble a rosary bead since they are sporadically dotted throughout the valley floor. Also known as ribbon lakes. See U-shaped valley, Glaciation.

Pearl Harbor

A sea inlet in Honolulu, Hawaii, that is the location of the U.S. naval power in the Pacific. In 1941, a surprise attack on Pearl Harbor by the Japanese forced the U.S. into World War II.

Peat

A fossil fuel that develops in marshy ground over millions of years from decayed vegetation and animal remains.

Peking

The capital city in China, also known as Beijing.

Peneplain

Land worn down by the agents of erosion until it is almost a flat plain.

Peninsula

A land area projecting out into the sea. Sometimes a land area almost completely surrounded by water and connected to the mainland by an isthmus. See also; Jutland, Iberia.

Pennsylvania

A state in eastern U.S., known as the keystone state of the original thirteen. It was founded by William Penn, a quaker from England who, together with other followers, wanted a place to practice their religious beliefs without fear of violence. Pennsylvania became a manufacturing and steel producing region because of its rich reserves of coal. The capital city is Harrisburg.

Permeable

A substance like a rock that is porous and can allow fluids, especially water, to penetrate and pass through. See also; Artesian basin.

Persian Gulf

An arm of the Arabian Sea between Iran and Arabia. The Gulf region has more than half the world's known oil and natural gas reserves, a fact that resulted in a great war, in 1991.

Peru

A country in western South America, on the Pacific Ocean. The capital city is Lima.

Petrified Forest

A national park in the Painted Desert of Northern Arizona containing the petrified trunks of several coniferous forests dating back to the Triassic period. This landscape was inundated and while underwater the tree trunks were transformed into stony substances but kept their original look and feel. As a result a twig may look like a fallen limb, but it might weigh several pounds.

Philippines

A country consisting of more than 7,000 islands in the southwestern Pacific Ocean. The capital city is Manila.

Photosynthesis

The process that occurs in green plants, when the leaves which contain chlorophyll use the energy in sunlight to combine carbon dioxide and water to make food. This is how plants release oxygen and use up carbon dioxide.

Physical map

A map that details the landscape of a region, showing mountains, valleys, rivers and so on.

Physical weathering

The process by which rocks are broken down and eroded by the natural elements, such as rain, ice, and waves.

Pisa

A town on the river Arno in northwestern Italy noted for its interesting leaning tower. It is a center for tourism.

Plain

A broad almost flat stretch of land, usually found along the lower reaches of rivers or along a coast.

Planet

A celestial body in orbit around a star, in space. The nine planets in our solar system orbit their star - the sun. Planets do not give off light, but they can reflect it.

Plate tectonics

The theory that the planet earth is made up of a number of solid plates that float on a molten bed and collide with one another causing earthquakes and vulcanicity.

Plucking

The action whereby a moving glacier breaks chunks of the bedrock away and removes it. This is a direct result of freeze-thaw action.

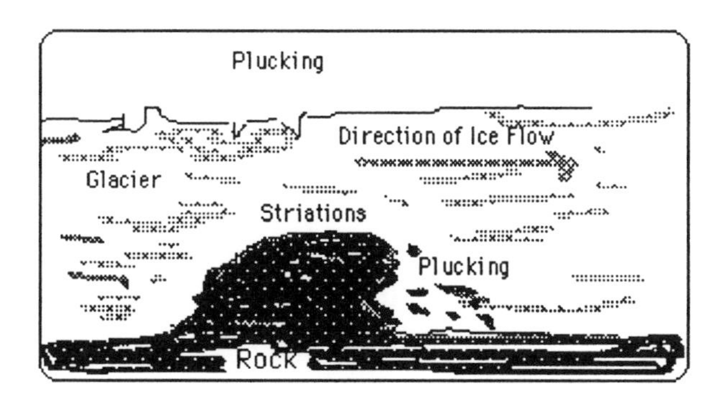

Fig. P i Plucking

Pluto

A planet, the furthermost one from the sun, in our solar system.

Pocono

A mountain ridge in eastern Pennsylvania.

Poland

A country in central Europe bordering on the Baltic. The capital city is Warsaw.

Polar front

A weather pattern caused by a front of Arctic cold air mixing with a warmer body of air and resulting in stormy weather.

Polar ice cap

The continental ice sheet that sits atop the polar regions of the globe.

Polaris

The north star that is almost directly above the North Pole. See also; North Star.

Political map

A map that describes the political relief of a country, usually with a color code showing different countries.

Pollution

The waste product of modern industry, or other harmful substances which are released into the air, rivers, sea or on land. It endangers human, animal, fish and plant life.

Polynesia

An island chain in the central Pacific including Hawaii, Phoenix, Tonga, Somoa, Cook and others. Polynesia means "many islands".

Pompeii

An ancient city in southern Italy that was destroyed by the erupting Mount Vesuvius in 79 AD, and many people were buried in ash and molten lava. Later the ruin was excavated to reveal the destruction, the petrified remains of the townspeople and, also, the well preserved artifacts that reveal much about the way of life in 79 AD.

Population

The total number of people who live in a country or region.

Portugal

A country of the Iberian peninsula, situated between Spain and the Atlantic Ocean. The capital city is Lisbon.

Precambrian

The earliest geologic period. Something that is exceedingly old.

Precipitation

Rain, hail, sleet or snow that derives from a cloud.

Prime Meridian

The imaginary line that circles the globe from pole to pole and passes through Greenwich. It is designated as the line of noon when the sun is directly overhead. Its importance lies in the fact that it is the first meridian and all others are numbered east and west, in angular degrees, from it. The line opposite the Prime Meridian is the International Date Line, at 180 degrees, and passes between America and the Soviet States through the Bering sea.

Ptolemaic Geocentric theory of Cosmology

The theory propounded by Ptolemy that the earth was the center of the universe and that the sun and all the planets revolved about it in large circular orbits.

Ptolemy

An astronomer and geographer who lived in the second century AD.

Puerto Rico

An island in the West Indies, east of Hispaniola and a commonwealth of the U.S. The capital city is San Juan.

Puget Sound

An inlet of the Pacific Ocean in the western U.S. Seattle, a busy seaport city of Washington State, is situated on Puget Sound.

Pumice

A volcanic ash.

Pygmy

A people of African origin.

Pyramidal peak

A peak that is caused by three or more glaciers eating back-to-back into the mountain removing large chunks of it. The Matterhorn is a prime example of a pyramidal peak. See also; Mountain - Ranges, Alps.

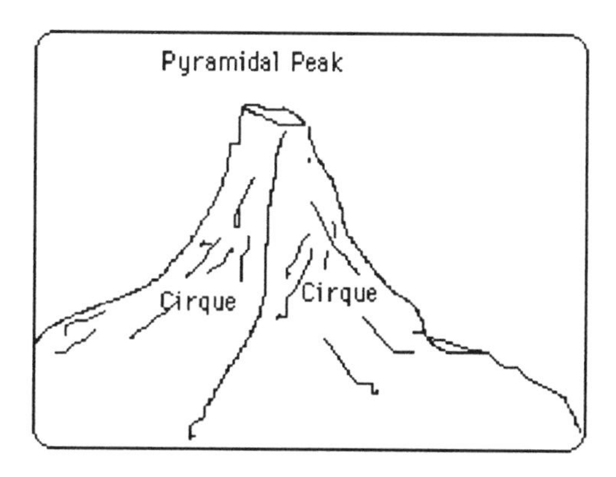

Fig. P ii Pyramidal Peak

Pyrenees

A mountain range along the French, Spanish border stretching from the Bay of Biscay to the Gulf of Lions.

Pyrite

A flint or millstone rock.

Qatar

A peninsular country in eastern Arabia projecting into the Persian Gulf. Qatar's economy is based on oil and it ranks as one of the richest countries in the world because of the huge reserves. The capital city is Doha.

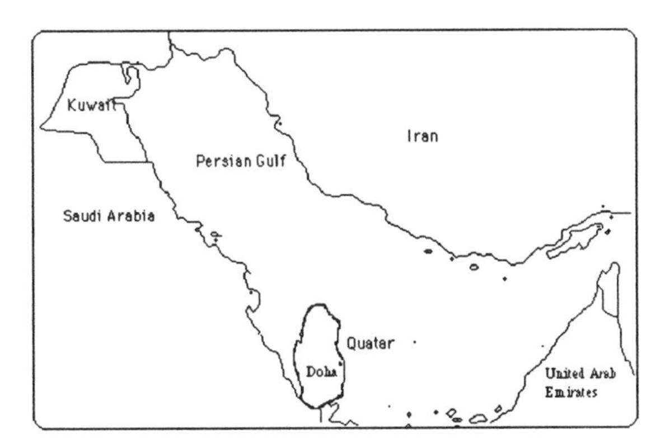

Fig. Q i Qatar

Quark

One of the fundamental building blocks of matter, coined from a phrase by James Joyce in *Finnegan's Wake.*

Quartz

The principal constituent of ordinary sand, but a mineral commonly found in rock formations.

Quasar

A celestial body that is extremely distant from the earth and that emits intense light and radio waves.

Quebec

The capital city of a Canadian province of the same name, extending from Hudson Bay to the Gaspe Peninsula. French is the principal language spoken in Quebec.

Queens

A borough of New York City on Long Island.

Queensland

The second largest state in Australia. The capital city is Brisbane.

Quito

The capital city of Ecuador in South America, situated in the Andes.

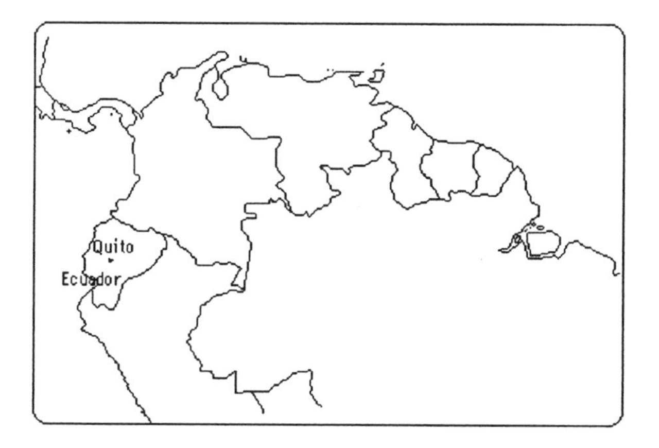

Fig. Q ii Quito

Radial drainage pattern

The river pattern that results when a number of drainage systems rise on the same upland area and flow in different directions to the sea.

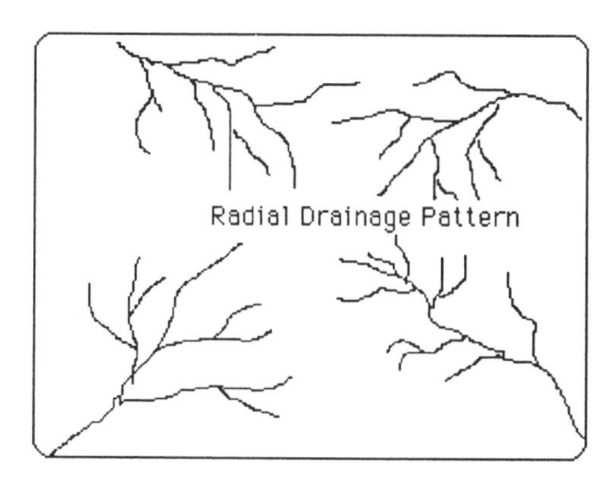

Fig. R i Radial Drainage Pattern

Radioactive

The property, possessed by uranium, of spontaneously emitting alpha or beta rays by the disintegration of the nuclei of atoms. In other words, as the atoms decay they release energy.

Rain

Water falling as drops condensed from vapor in the atmosphere.

Rain Forest

Dense forest growth that occurs because of the high rainfall and hot or mild conditions. Tropical rain forests occur near the equator where the weather is hot and moist throughout the year. Hardwoods, like ebony and mahogany, are produced in these forests and natural rubber can be harvested.

Rainbow

A colorful phenomenon that occurs in the sky after a shower when the sun is shining. The rain drops act as thousands of tiny prisms and mirrors breaking the sunlight into the colors of the spectrum and reflecting them back to the observer.

Rain gauge

An instrument for measuring the quantity of precipitation. It resembles a cylinder that is calibrated and has sloping, sharp lips that cut drops so that accuracy is achieved and evaporation is curtailed.

Raised beach

Due to isostasy in post glacial periods the sea level rises and falls in relation to the land and beaches are sometimes left high and dry above the present sea level.

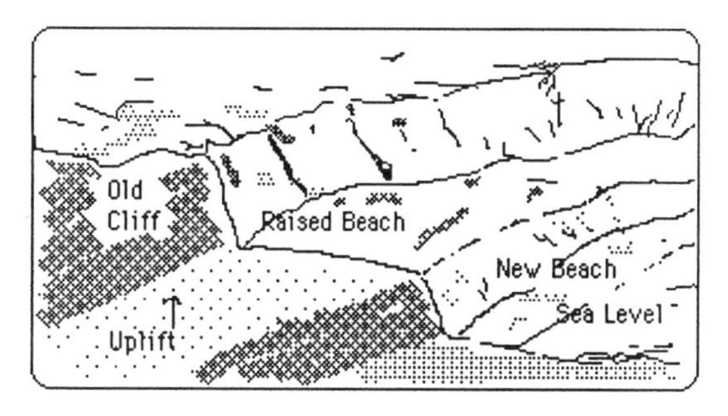

Fig. R ii Raised Beach

Rapids

Part of the river, usually in the upper course when the flow is turbulent and the water has to overcome many obstructions.

Recession

A time of reduced economic activity.

Recessional moraine

The depositional mound of boulder clay left in the wake of the retreating glacier.

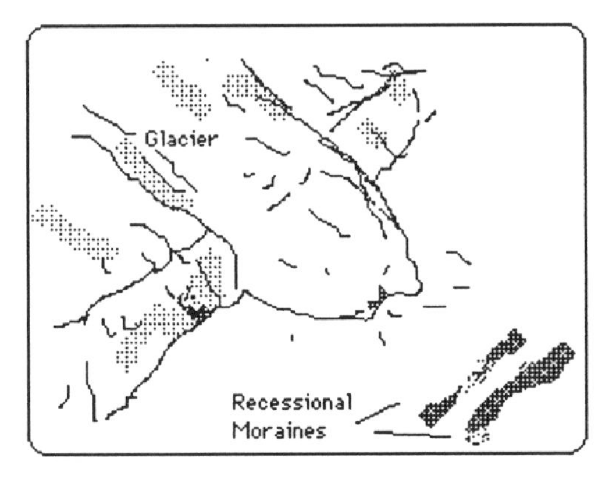

Fig. R iii Recessional Moraines

Recycling

To re-use the natural resources again in an effort to conserve.

Red Sea

A long narrow inlet of the Indian Ocean, that separates Saudi Arabia from northeastern Africa. The Red Sea is now one of the busiest waterways in the world since the Suez Canal was opened, connecting it to the Mediterranean Sea.

Red-shift

The change in the wavelength of light when the source is moving away from the observer.

Reef

A chain of rocks near the surface of the sea, e.g. Barrier Reef.

Region

A broad homogenous geographical area, e.g. Great Plains, Rocky Mountains.

Relative humidity

The ratio of the amount of water vapor actually present in the air to the greatest amount possible at the same temperature. A relative humidity of 50% means that the air is holding only half the amount of water possible.

Relief

The elevations of a land surface, the mountains and the valleys.

Representative fraction

A type of scale used on a map that uses mathematical formulae to overcome national and linguistic barriers.

Retreat - ice

When the melt is greater than the snowfall, then the glacier retreats. Opposite to advance.

Reykjavik

The capital city and chief port of Iceland.

Rhine

A river in western Europe that rises in Switzerland, flows through Germany and enters the North Sea, at Rotterdam in the Netherlands. It is the most important inland waterway in Europe.

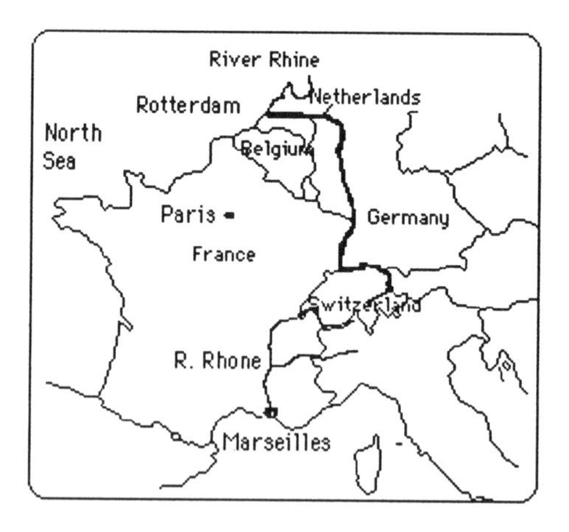

Fig. R iv River Rhine

Rhode Island

The smallest state in the U.S. Its capital city is Providence.

Rhodesia

A former South African country. See also; Zimbabwe.

Rhone

A river in western Europe that rises in Switzerland and flows south through France and empties into the Mediterranean at Marseilles. It is an important inland waterway in France.

Ria

A drowned river valley, due to post-glacial isostatic movement. Rias differ from fjords since they have sloping sides, as opposed to vertical drop-offs. See also; Headland.

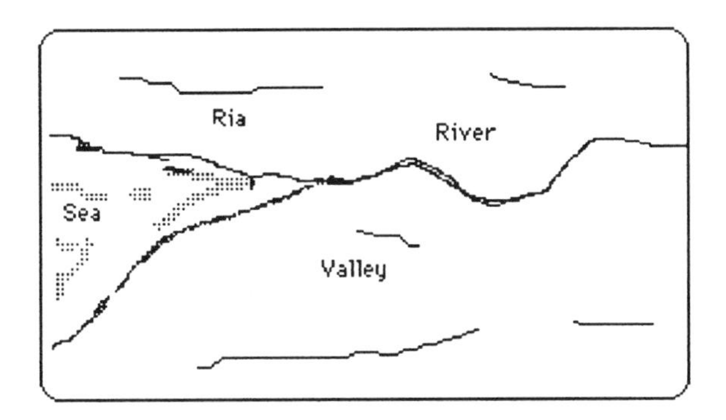

Fig. R v Ria

Ribbon lakes

The lakes that dot a glaciated U-shaped valley. See also; Pater noster lakes.

Richter scale

A scale that expresses the magnitude of a seismic disturbance like an earthquake. A measurement of more than six is very devastating to the landscape affected.

Rift valley

A large tract of land that forms a valley when a block of terrain falls between two faults. It is also known as Graben. A great rift valley runs for thousands of miles from northeastern Africa to Madagascar.

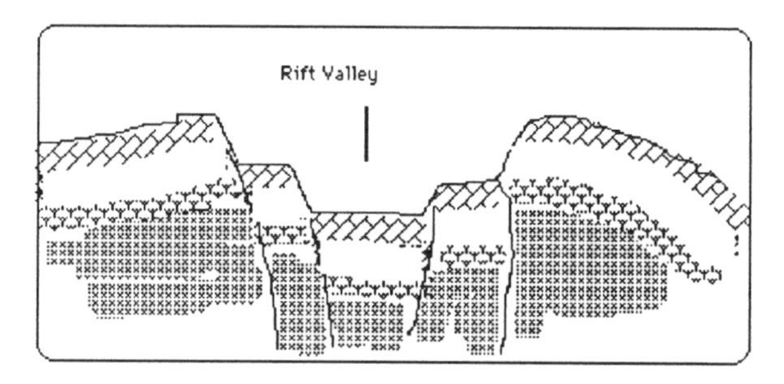

Fig. R vi Rift Valley

Rio de Janeiro

A large city, port and tourist center in southeastern Brazil. It was formerly the capital city.

Rio Grande

A river in southwestern U.S. forming part of the border with Mexico. It flows from the Rocky Mountains in Colorado to the Gulf of Mexico.

River

A large stream of water that occupies a watercourse or valley.

- alluvial fan

When a river slows down or stops upon entering the sea or a lake it drops its load in a peculiar fan shaped heap called alluvial fan.

- braided stream

In the upper course a river will break into a number of channels to negotiate shallow sections in its bed and later rejoin in deep pools. These rivulets are referred to as braided streams.

- interlocking spurs

In the upper course the valley is V-shaped and the spurs appear to be interlocking when one looks up river toward the source.

- rapids

In its upper course, a river invariably tumbles over rocky outcrops. Differential erosion forms a series of small falls, known as rapids.

- rejuvenation

When isostatic recoil occurs in the wake of glaciation a river can begin down-cutting again, within its original channel, to reach the new lower sea level.

- ria

In the aftermath of glaciation when isostasy causes the sea level to rise relative to the surrounding landscape, many low-lying areas are drowned. Extant V-shaped, river valleys become rias. U-shaped glaciated valleys become fiords.

Riviera

A tourist coastal region on the Mediterranean Sea shared by France and Italy.

Riyadh

The capital city of Saudi Arabia.

Roches moutonnee

This is an elongated rounded ice-sculptured chunk of bedrock. It resembles a French wig, that was worn by judges in past times. It is a result of the polishing and scouring effect of the ice on the front side of the rock and the plucking action on the leeward side of the rock. See also; Plucking, Glaciation.

Rock

A solid mass of mineral matter that makes up the hard outer layer of the earth. Rocks can be either igneous, sedimentary or metamorphic, according to their origin.

Rock-flour

The material that is left as a result of the disintegration of rock by ice and its subsequent conversion to soil.

Rocky Mountains

A mountain chain that forms the backbone of the western U.S.

Romania

A country in eastern Europe. The capital city is Bucharest.

Rome

The capital city of Italy, and a center of ancient history and culture.

Runoff

Water that falls as precipitation and eventually ends up as groundwater. It is called runoff as it finds its way into the river channels and lakes.

Rushmore, Mount

A mountain in the Black Hills of South Dakota, upon which are sculpted giant carvings of the faces of Washington, Jefferson, Lincoln and Roosevelt. It is a National Memorial and a favored tourist attraction.

Russia

A member country of the Commonwealth of Independent States located on the Asian continent. The capital city is Moscow.

Sacramento

The capital city of California, situated on a river of the same name.

Sahara

A desert region in north Africa, that stretches from the Atlantic to the Red Sea. Approximately, equal in area to the USA, it is the world's largest desert.

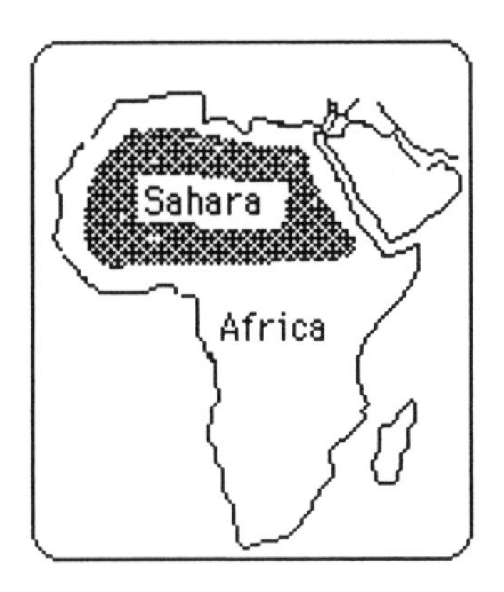

Fig. S i Sahara

Sahel

A region in Africa, south of the Sahara, that is characterized by severe droughts.

Saint Helens, Mount

An active volcanic mountain in Washington state. Site of a spectacular eruption in 1980.

Saint Lawrence

A river in Ontario, Canada, that borders with the U.S. and affords a strategic navigable deep water channel between Lake Ontario and the Atlantic via the St. Lawrence Seaway.

Saint Lawrence Seaway

A system of canals and dams between Lake Ontario and Montreal that permits the passage of ocean-going vessels for commerce.

Salt Lake City

Capital city of Utah, in western U.S.

Saltation

A method used by rivers to bounce heavy material along in their flow.

San Andreas Fault

A major fault line dividing the Pacific plate from the North American plate. Earthquakes and vulcanicity take place on a frequent and sporadic basis as a result of the movement of the plates along the fault and the strain it produces on the crust.

San Diego

A city in southern California and a major center of naval and aerospace activity.

San Francisco

A city and port in western California, situated on a bay of the same name. San Francisco is a center of American culture and a major tourist attraction.

Fig. S ii San Francisco

San Joaquin

A river valley in central California, drained by the river of the same name. Once a desert area it was irrigated and is now a region of intensive agriculture that produces much of the fruit and vegetables for the U.S.

Sand

Loose granular material that results from the disintegration of rocks. It is common on shores of seas and rivers and some deserts. Also the ingredient of glass.

Sand Bar

A sand deposit resulting from long-shore drift or currents, that causes an obstruction in estuaries and river mouths.

Sandstone

A sedimentary rock formed by material laid down by rivers and seas over millions of years on beach areas. It consists of sand and other sediments cemented together by silica.

Santa Fe

The capital city of New Mexico.

Sargasso Sea

A region of the North Atlantic near Bermuda where the sea is forced into a circular motion and is comparatively slow at the center, due to the configuration of the eastern coast of North America and the Gulf of Mexico. The ocean current that drifts into the North Atlantic circles back around Bermuda and sea weed collects in this area. The sea at this point is rich in marine life and displays a deep blue color.

Sardinia

An Italian island off the Italian coast south of Corsica. Its capital is Cagliari.

Saskatchewan

A province in the west of Canada that has a rich lacustrine soil and is suitable for the production of cereal crops. Regina is its capital. Also a river that rises in the Rockies of western Canada and flows east into Lake Winnipeg, Manitoba.

Satellite

A celestial body, e.g. a moon, that revolves around a larger body or planet. Also an orbiting craft deployed from earth to send back information relating to the earth or the other planets.

Saturn

The planet of our Solar System that is sixth from the sun.

Saudi Arabia

A country in the Middle East made up mostly of desert and semi-desert land. It is one of the world's leading producers of petroleum. The capital city is Riyadh.

Scale

A measure of the distance on a map in relation to the actual distance on the ground. There are three common representations of scale in use for mapping purposes. A linear, a graphic scale and a representative fraction.

- Large-scale

A large scale map displays a small amount of territory, but in good detail. A scale of one inch to ten feet will not cover much ground but one can get a great deal of detail in a map drawn to this scale.

- Small-scale

A small scale map will display a large amount of territory with less detail. This way one can get a view of a large area like the world but without much detail.

Scandinavia

The name given to a group of countries in northern Europe - Denmark, Norway and Sweden.

Scotland

A country north of England in the United Kingdom. Its capital city is Edinburgh.

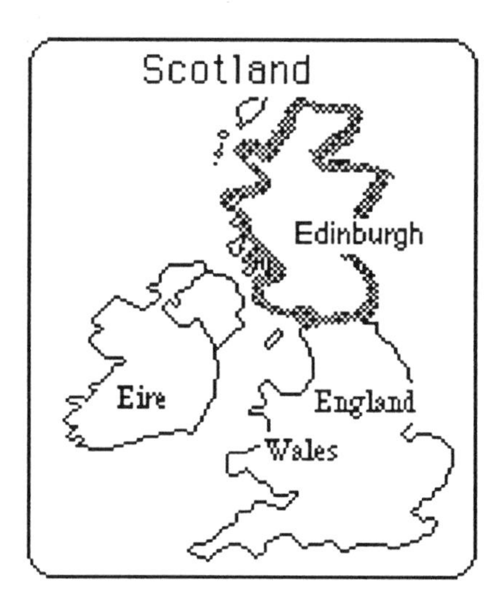

Fig. S iii Scotland

Sea-arch

A rock structure resulting from wave erosion on a headland from both sides. When caves are eroded back to back they sometimes join to form a sea arch.

Sea-Arch

Sea breeze

A wind that flows from the sea to shore, usually during the day. Land heats up faster than water and the warm air rises and is displaced by cooler sea air.

Sea-level

The level of the surface of the sea between high and low tide. Geographers use average sea-level as the starting point for measuring the elevation of places on the earth.

Sea-stack

A rock formation resulting from differential wave erosion seen usually on the coast. See also; Sea-arch.

Sea-Stack

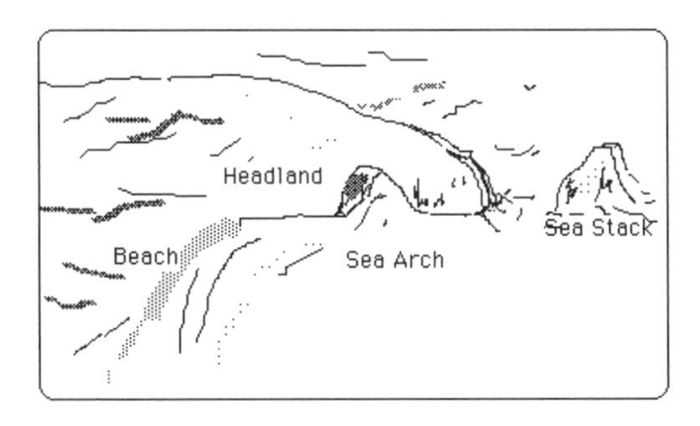

Fig. S iv Sea-Arch, Sea Stack

Season

One of the four quarters into which the year can be divided: Spring, Summer, Autumn, Winter.

Seattle

A city and port in Washington State, on the west coast of North America. See also; Puget Sound.

Sedimentary rock

The rock strata that is laid down layer upon layer over millions of years usually at the bottom of the sea or a large lake. It is cemented together to form a layered rock. Fossils can be trapped in these layers, providing keys to the geologic and plant activity of an era.

Seismic waves

The shock waves that travel out from the epicenter in the aftermath of a powerful earthquake.

Seismograph

An instrument for measuring seismic activity.

Senegal

A country on the coast of northwestern Africa. The capital city is Dakar.

Seoul

The capital city of South Korea.

Serac

An ice edifice among the crevasses on a glacier.

Seychelles

An African island country in the western Indian Ocean, noted for tourism because of its pleasant climate and attractive beaches. The capital city is Victoria.

Shanghai

A major port and industrial center in China and one of the largest cities in the world.

Shale

A sedimentary rock that is very finely stratified and consists mostly of cemented mud and silt.

Shenandoah

A river flowing northeast between the Allegheny and the Blue Ridge mountains into the Potomac in Virginia.

Shetland

The islands about one hundred miles north of Scotland beyond the Orkney islands, famous for their ponies and a rugged brand of sheep from which woolen garments are manufactured by local crafts-people. Many people make a living from part-time farming and part-time fishing.

Shooting star

A streak of light in the night sky caused by a meteor falling through the atmosphere and burning up.

Shoreline

The area between the land and the tides.

Siberia

Remote region of Russia east of the Urals.

Sicily

An island off the toe of Italy, in the Mediterranean sea. The capital city is Palermo. See also; Italy.

Sierra Leone

A small country in West Africa, just north of the Equator. It is one of the chief producers of diamonds. Freetown is the capital city.

Sierra Nevada

A mountain chain in California.

Sill

An intrusive volcanic plug, made of lava, that penetrates an horizontal weakness in a rock stratum. See also; Volcano, Crater.

Silt

The fine deposit laid down in lakes and rivers. See also; Alluvium, Alluvial Fan.

Silver

A very valuable white metallic element that has many fine uses for jewelry manufacturing and industry.

Singapore

The capital city of an island country of the same name, in the Malay archipelago.

Sink hole

In limestone topography a hollow that connects with underground drainage patterns.

Slate

A dense fine grained metamorphic rock consisting of mud and silt. It can easily be cleaved so that it produces thin sheets with a smooth water-resistant finish. It is used in many countries for roofing and as a paving stone.

Sleet

A form of precipitation that is a mixture of snow and rain.

Slip-off slope

A river feature at a meander loop where undercutting is occurring on one side and deposition is happening on the other. The side with the deposition is a slip-off slope.

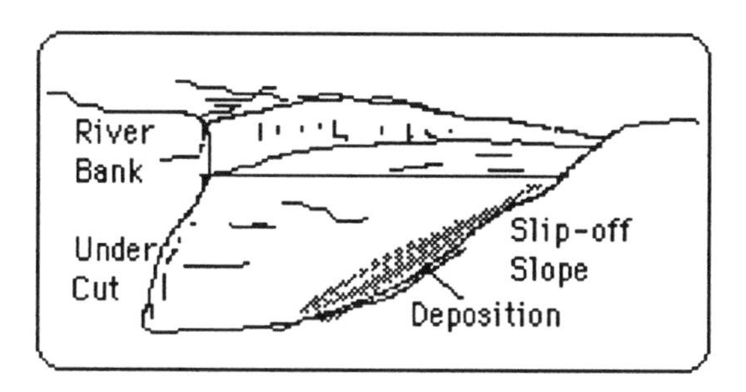

Fig. S v Slip-off Slope

Slovak Republic

A new republic in the division of the former Czechoslovakia. The capital city is Bratislava.

Slump

After rainfall, solifluction causes soaked chunks of land to slide or slump because of the lubricating effect of the moisture.

Smog

A particular atmospheric condition caused by the filling up of the air with pollutants, especially exhaust fumes in an inversion layer. The word derives from a combination of smoke and fog, in an attempt to describe the air.

Snake

A legless, often venomous reptile with a long tapering body.

Snake, River

A tributary to the Columbia river. This picturesque river rises in Yellowstone National Park in Wyoming and flows through the Grand Tetons National Park. It passes through Hells Canyon in Idaho, before entering Oregon and Washington where it joins the Columbia.

Snout

The front of a glacier. If the glacier is at the coast, parts of the snout often slump into the sea, forming icebergs. See also; Glaciation.

Snow

A form of precipitation that occurs when it is cold enough for water vapor to form ice crystals in the clouds.

Snow Line

The lower limit of upland that is permanently covered with snow.

Soil

A layer of mineral and decomposed organic material that covers much of the surface of the earth and in which vegetation grows.

Soil erosion

The weathering, erosion and transportation that occurs on the earth's surface, by agents of the elements, wind, ice, water and gravity. Soil can be blown away by wind, washed away by rain and rivers and transported away by ice. Many agricultural and lumbering techniques lead to soil erosion also.

Solar eclipse

The obscuring of the sun by the moon as it passes between it and the earth.

Solar orbit

An orbit that takes a celestial body around the sun.

Solar power

That kind of energy that is produced by making use of the heat and light of the sun.

Solar Radiation

The sun's energy transported throughout the universe by radiation.

Solar System

The nine planets including the earth that orbit about the sun.

Solar wind

The wind, a continuous flow of gases, that flows from the sun causing tails to occur on comets.

Solifluction

The seepage of water into the soil causing it to become lubricated to the point where slumping can occur.

Solstice

The term given to two days during the year when the earth's axis is tilted at its greatest angle toward or away from the sun. The summer solstice is generally on June 21 and the winter solstice falls usually on December 22.

Somalia

A country at the horn of Africa. It is often stricken by famine since it is located in an arid and harsh climatic zone. The capital city is Mogadishu.

Source

The place where a river or a stream rises or begins.

South Africa

A country at the tip of southern Africa. It is the richest country in Africa, renowned for its gold and diamond production. The legislative capital is Capetown.

South America

A continent in the southhern hemisphere that is connected to North America by Central America.

Fig. S v South America

South Australia

A state on the southern coast of Australia. The capital city is Adelaide.

South Carolina

One of the original thirteen states of the U.S. The capital city is Columbia.

South Dakota

A farming state in the Midwest. The capital city is Pierre.

South East Asia

A region to the southeast of China, that includes Burma, Thailand, Laos, Vietnam, Kampuchea and many neighboring island countries.

South Equatorial current

The warm body of water that originates at the equator and flows south to mix with the colder currents that come from the Antarctic.

South Pole

The southernmost point of the earth, located in Antarctica.

South Sea Islands

A chain of islands in the south Pacific, also called the Pacific Islands/Oceania.

Space

The region beyond the earth's atmosphere.

- and time

Einsteinian theories relating to relativity in which light and time can be bent by gravitational effects of large bodies in space.

- exploration

A commitment to discover what exists beyond the pull of earth's gravity.

- race

A phenomenon that occurred when a Russian space craft, Sputnik, beat American efforts into space.

- Shuttle

The ship that leaves the earth's atmosphere in order to carry out missions in outer space and can re-enter to be prepared for further flights.

- age

The age that was born since the launching of Sputnik in 1957.

Spain

A country of western Europe in the Iberian peninsula. Madrid is the capital city. See also; Iberia.

Spectral analysis

The study of the spectrum of visible light.

Speed of light

The velocity at which light travels.

Spit

A depositional feature along the coast where sand bars are laid down and configured by long-shore drift and currents so that they extend from the land into a channel.

Spring

The first of the four seasons of the year marked by the sun's overhead appearance on March 22nd at the Tropic of Cancer in the Northern Hemisphere.

Sputnik

The USSR satellite that was the first to enter orbit around the earth in 1957.

Sri Lanka

An island country in the Indian Ocean, formerly called Ceylon. The capital city is Colombo.

Stalactites

In limestone topography, the build-up of calcite on the roofs of caverns caused by the solidifying of calcium-rich droplets that evaporate before they fall to the floor below.

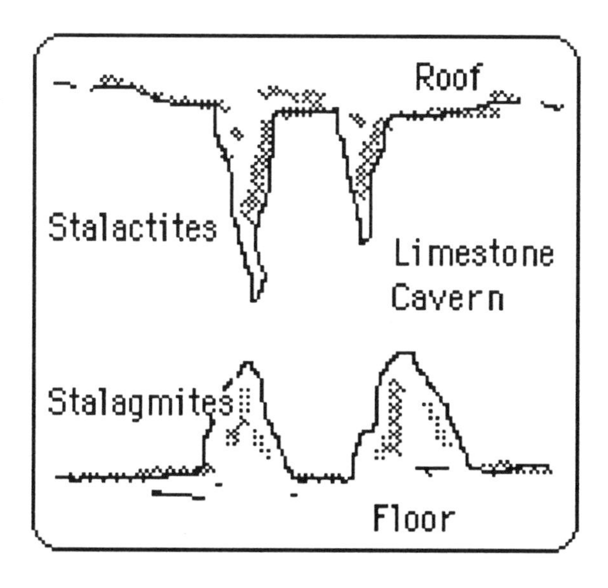

Fig. S vi Stalactites

Stalagmites

In limestone topography the build-up on the floors of caverns caused by the solidifying of calcium-rich droplets from the walls and roof.

Stonehenge

A formation of megaliths on the Salsbury Plain, Wiltshire, England, erected by a prehistoric people. A major tourist attraction today.

Stockholm

The capital city of Sweden, situated on the Baltic Sea.

Stratosphere

The upper portion of the earth's atmosphere.

Stratus cloud

High air masses that are stratified, and appear smooth from below.

Striation

In glacial topography the scratches that appear on rocks that can indicate the direction that the glacier moved. See also; Glaciation.

Strip grazing

A form of agriculture where technological controls, usually electric fences, contain livestock to particular areas of grazing so that other portions of the field are allowed to flourish.

- mining

Intensive mining of ores and natural resources usually by open cast method. It causes destruction to the environment and to the physical landscape.

Sub-tropical climes

Areas near the equator that experience warm climates.

Subduction zone

The zone at the edge of plate tectonic activity where one plate is forced under another and volcanic activity usually results.

Submarine contours

Imaginary lines on a map that join places of the same depth below sea level.

Sudan

A desert country, located south of Egypt, in northeastern Africa. The capital city is Khartoum.

Suez Canal

A canal that connects the Red Sea to the Mediterranean Sea. Ocean-going vessels no longer have to round the tip of Africa as they conduct commerce between Europe and the Far East.

Summer Solstice

The time of year when the sun is farthest from the equator. (June 21, 22 in the northern hemisphere). See also; Solstice.

Sun

A star - the center of our solar system.

Sun dial

An instrument that indicates time by the position of a shadow cast by a gnomon on a calibrated face.

Sun spot

A cool region appearing sporadically on the face of the sun caused by geomagnetic disturbances.

Supernova

An extremely bright star that suddenly increases in brightness by millions of times.

Suspension

A river carries its load of sediment and detritus by suspension when it is dissolved in it. Typically the river will take on a characteristic muddy color when there is a lot of silt in it.

Swash

The wave crashing on to the shore runs up the beach as swash and falls back as backwash.

Swaziland

A small landlocked country in southern Africa. The capital city is Mbabane.

Sweden

A country in northern Europe, part of the Scandinavian peninsula west of the Baltic sea. Stockholm is its capital city.

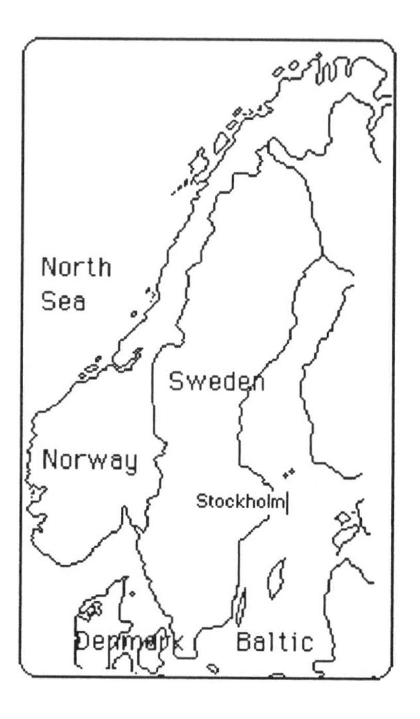

Fig. S vii Sweden

Swell

The agitated sea after a storm, with rolling breakers and surging waves.

Switzerland

A country in central Europe situated in the Alps. Bern is its capital city. Because of its tradition of neutrality it is one of the world's leading financial and banking centers. Switzerland is surrounded by France, Germany and Italy and as a result many inhabitants can speak a number of European languages.

Sydney

The capital city of New South Wales, Australia.

Symbols

On a map symbols are used to note the scaled items that represent the real world.

Syncline

A down-fold in topography, opposite to anticline.

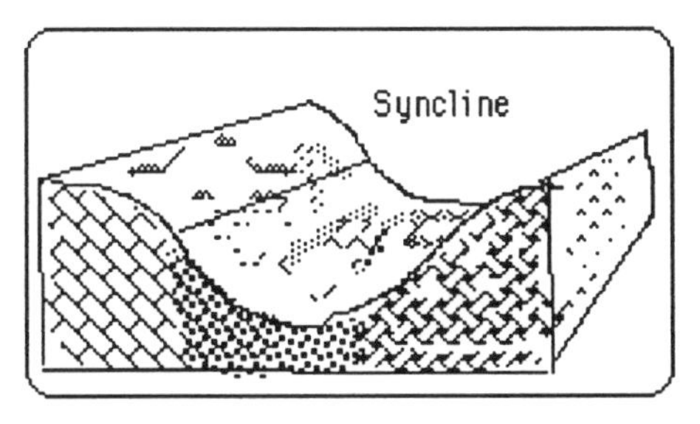

Fig. S viii Syncline

Syria

A country in southwest Asia, bordering on the Mediterranean. The capital city is Damascus.

Tahiti

An island in the South Pacific, made famous by Gaugain for its beauty. The capital city is Papeete and is located on Tahiti, the largest island in French Polynesia.

Tahoe

A lake and popular tourist attraction along the California, Nevada border.

Taiwan

An island country in the Pacific, east of China. The capital city is Taipei.

Taipei

The capital city of Taiwan.

Talus

Also known as scree, it is a mantle of rock fragments at the base of a slope or cliff.

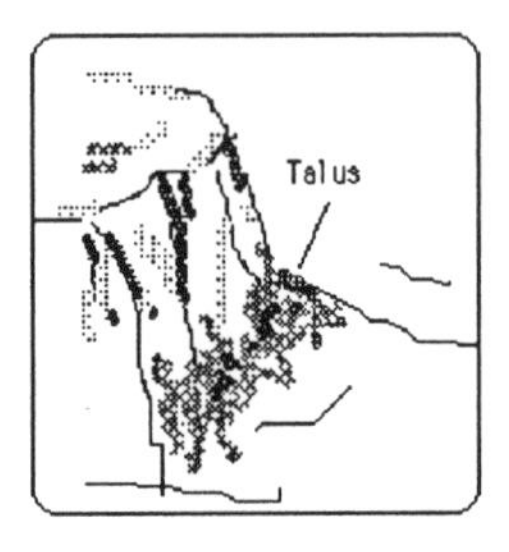

Fig. T i Talus

Tanzania

A republic in East Africa. The capital city is Dar es Salaam.

Tara

A village in County Meath, Ireland. It was once the seat of the ancient Irish kings.

Tarn

A deep lake in glaciated landscape, sometimes called a corrie or a coom.

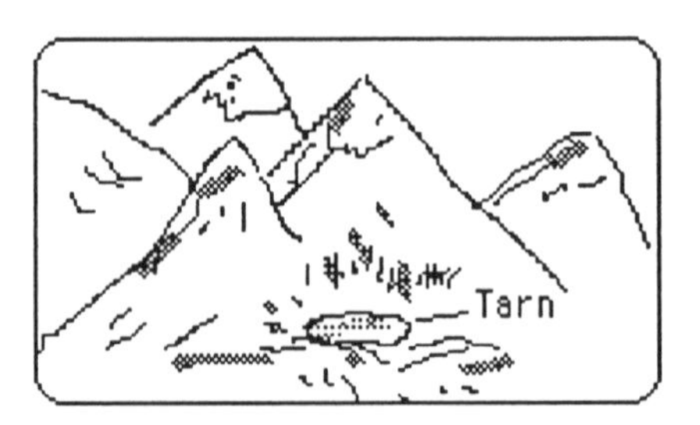

Fig. T ii Tarn

Tasmania

An island state off the south coast of Australia. The capital city is Hobart.

Tectonic plate

The units of floating lithosphere on the molten magma that make up the surface/crust of the earth.

Teheran

The capital city of Iran.

Telescope

An instrument used for magnifying far-away places and is especially practical for stellar observations.

Temperature

The measurement of how hot or cold a place or object is.

Tennessee

A state in eastern U.S. The capital city is Nashville. Also a tributary to the Ohio river.

Tennessee Valley Authority

The governmental body, created to conserve the resources of the Tennessee valley region. It was responsible for development of the upper tributaries of the Mississippi valleys and, thereby, controlled flooding in the lower course.

Terminal moraine

The final moraine that is deposited by the retreating glacier.

Terra Incognita

On Ptolemy's early maps this was the name he gave to the undiscovered parts of the earth. It means "unknown land".

Terrace

In river topography a terrace is a raised platform where the river valley used to reside before the latest migration and down-cutting.

Texas

A state in southwestern U.S. Texas is second only to Alaska in size. The capital city is Austin.

Thailand

A tropical country in Southeast Asia, southeast of Burma. The capital city is Bangkok.

Thames

A river in the south of England upon which London is situated.

Fig. T iii Thames

Thematic map

A map that focuses on a particular theme, like rainfall amounts.

Third world

The developing countries of the earth where economies are generally based on agriculture.

Three Mile Island

The site of a nuclear power plant near Harrisburg, Pennsylvania, where a serious accident occurred in 1979.

Throw-away society

A name given to modern society before recycling became popular.

Thunderhead - cloud

A particular shape to a stormy pattern of clouds when warm air is suddenly lifted into higher altitudes.

Tiber

A river in central Italy that flows through Rome westward into the Tyrrhenian Sea.

Tibet

A high and remote plateau region in south central Asia. The inhabitants are noted for their simple, primitive ways and religious fervor. The capital city is Lhasa.

Tidal bore

In narrow estuaries the difference between low and high tides can be very large and is referred to as a tidal bore. See also; Bay of Fundy.

Tidal reaches

The area of land between high and low tides that is susceptible to periodic inundations.

Tidal wave

An unusually catastrophic wave that

results from earthquakes or tectonic movement. See also; Tsunami.

Tides

The effect of the moon on the sea areas so that there is a rise and fall of ocean water every day. This is caused by the gravitational impact of the moon on all earthly bodies.

Tierra del Fuego

An archipelago off the southern tip of South America in the straits of Magellan divided between Argentina and Chile.

Fig. T iv Tierra del Fuego

Tigris

A river that runs through southern Turkey and Iraq and flows south-southeast to its confluence with the Euphrates.

Till

Also known as glacial till, or boulder clay, this is the rich arable soil that develops in some instances in the aftermath of a period of glaciation.

Timbuktu

A town in Mali, West Africa near the river Niger.

Time zones

The earth is divided into different time zones for convenience, since the sun can shine directly on only one place at a time.

Titicaca

A lake along the Bolivia, Peru border, which at 12,500 feet, is the highest lake in the world.

Tokyo

The capital city of Japan and one of the largest cities in the world.

Tonga

An island country in the South Pacific. The capital city is Nukualofa.

Top soil

The A horizon or humus, where minerals and moisture allow crops and plants to grow. See also; Horizon.

Topography

The physical landscape.

Tornado

A stormy wind pattern caused by differential pressures and temperatures of air masses. Tornadoes can be very destructive.

Toronto

The capital city of Ontario, Canada.

Tourism

An industry in economic geography where travel and lodging is turned into a way of life.

Trade wind

The prevailing easterly wind in the tropics, caused by the Coriolis force.

Transporting - work of rivers

When the river erodes the sides and banks of its channel the material is transported in suspension, by saltation or by hydraulic action.

Transporting - Economic geography

The movement of people and goods from one location to another for commercial and economic purposes.

Trellised drainage pattern

The river pattern that develops on karst or limestone topography where angular shapes arise. See also; Karst.

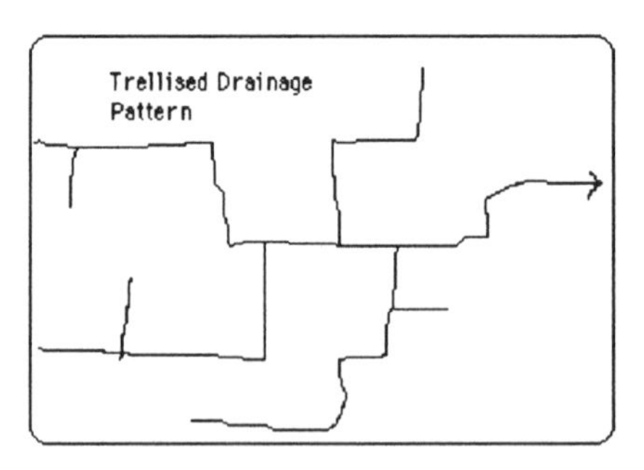

Fig. T v Trellised Drainage Pattern

Tributaries

A streamlet that joins with others and with the main branch to form a river.

Trinidad

A country that consists of two islands in the Caribbean off the coast of Venezuela. The capital city is Port of Spain.

Tripoli

The capital city of Libya.

Tropic of Cancer

The imaginary line on the globe 23.5 degrees north of the equator marking the northernmost path of the sun.

Tropic of Capricorn

The imaginary line on the globe 23.5 degrees south of the equator marking the southernmost path of the sun.

Tropics

The regions that lie between 23.5' north and south of the equator and where the sun constantly shines overhead. These regions are characterized by higher temperatures than the rest of the world.

Tropical Rain Forest

Lush forest growth in tropical regions of the world. See also; Rain forest.

Trough

The depression between two waves, as in the ocean or on a thematic weather chart.

Truncated spurs

The formation of a u-shaped valley in glaciated topography means that pre-existing spurs have to be truncated (cut-off) by the glacier to produce the characteristic steep sided valley, sometimes with accompanying waterfalls and ribbon lakes.

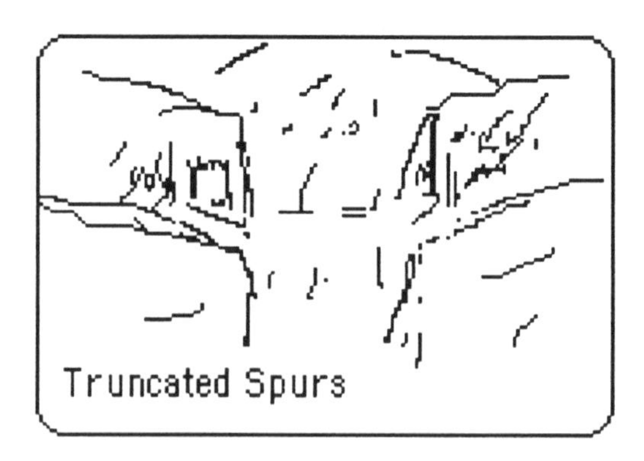

Fig. T vi Truncated Spurs

Tsunami

Another name for a tidal wave that usually results after an earthquake or tectonic movement. See also; Tidal wave.

Tundra

A cold, dry, treeless region caused by low temperatures and periglaciation either in high latitudes or high altitudes. Only the top three inches of soil will thaw periodically, allowing shallow rooted plants and mosses to colonize the tundra region.

Tuff

A volcanic rock.

Turkey

A country that lies partly in Europe and partly in Asia and between the Black Sea and the Mediterranean. Istanbul, the largest city, is located in Thrace on the European side. Ankara, the capital city, is in Anatolia/Asia Minor on the Asian side. See also; Istanbul.

Twilight

The period of semi-darkness at sunset.

Tyrannosaurus Rex (Dinosaur)

One of the large prehistoric dinosaurs.

Tyrol

A mountain region in the European Alps, particularly in Austria.

Uganda

A country in East Africa, north of Lake Victoria. The capital city is Kampala.

Ukraine

A republic on the north coast of the Black Sea in eastern Europe. Its capital city is Kiev. The Ukraine is a fertile tract of arable land where wheat, barley, corn and rye are grown in vast quantities.

Ulan Bator

The capital city of Mongolia.

Ulster

The northern province of Ireland, comprising nine counties, six of which are located in British-ruled Northern Ireland. See also; Ireland.

Ultraviolet

The invisible rays situated beyond violet in the visible spectrum of light. Ultraviolet rays from the sun can be dangerous to humans. They cause sunburn that may lead to skin cancer.

Unconformity

A lack of continuity in deposition in rock strata corresponding to a period of weathering or erosion.

UNESCO

United Nations Educational and Scientific and Cultural Organization. An agency designed to promote cooperation, respect and understanding among people everywhere.

UNICEF

United Nations International Children's Emergency Fund. An agency set up to help children everywhere to fight problems of hunger, disease and injustice.

United Arab Emirates

A federation of seven states on the eastern coast of the Arabian peninsula.

The economy is based on petroleum and natural gas. The capital city is Abu Dhabi.

United Nations

An organization that works for world peace. The headquarters are in New York.

Universe

The entirety of the celestial bodies.

Upper course

The portion of a river valley when the gradient is steep, the valley is marked by interlocking spurs and the work is mostly erosional.

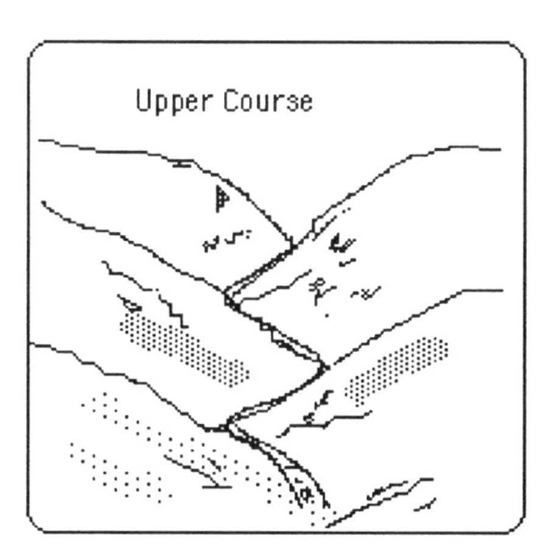

Fig. U i Upper Course

Uranium

A heavy radioactive metallic element used in the creation of energy by nuclear means.

Uruguay

A country on the east coast of South America. The capital city is Montevideo.

Ural Mountains

A mountain range that stretches from the Arctic circle to the Caspian Sea and divides Europe and Asia.

Uranus

The planet, seventh in distance from the sun, in our solar system.

Ursa Major

A constellation called the Greater Bear. It is conspicuous in the northern constellations, because it contains the Big Dipper with the pointers that look to the North Star.

- Minor

A constellation also called the Lesser Bear that contains the North Star and the Little Dipper.

U-shape

The overall shape of a glaciated valley with steep sides and a flat floor. See also; Glaciated U-Shaped Valley.

Utah

A state in western U.S. Its capital city is Salt Lake City.

Uzbekistan

One of the states in the Commonwealth of Independent States. The capital city is Tashkent.

V-shape

The over-riding shape of a river valley in its youthful stage. See also; Upper course.

Valley

The landscape formation that results from river erosion over time. A trough in the surface of the earth or on the sea floor, caused by block movement.

- U-shaped

The overall shape of a valley that has experienced glaciation. It has steep sides and a flat floor. See also; Glaciated U-shaped valley.

V-Shaped Valley

Vancouver

A city and sea port in southwest Canada in British Columbia. Also an island off the southwest coast of Canada. Its chief city is Victoria.

Venezuela

A country in South America, whose capital is Caracas. The economy is based largely on petroleum.

Venice

A city of northern Italy. This city has wonderful waterways that thread throughout its length and breadth and are plied by skillful gondoliers.

Fig. V i Venice

Vent

The outlet for magma in a volcano.

Ventifact

A feature that is formed by aeolian (wind) erosion, usually in desert landscapes.

Venus

The second planet from the sun in our solar system and the closest planet to earth.

Vermont

A rural and picturesque New England state. The capital city is Montpelier.

Vernal equinox

The Spring equinox that occurs on March 22 in the northern hemisphere, when the sun is directly overhead at the Equator.

Vertical erosion

The downward erosion that occurs in the bed of the river by hydraulic action and abrasion.

Vesuvius

A famous volcano in Italy on the Bay of Naples. It was responsible for the burial of Pompeii in 79 A.D.

Victoria

A state in southeastern Australia. The capital city is Melbourne.

Victoria Falls

A spectacular waterfall on the Zambezi river in South Africa, approximately 350 feet high and a mile wide.

Vienna

A wonderful old world city on the river Danube. It is the capital city of Austria and home of the Viennese waltz.

Vietnam

A country in southeast Asia, south of China. The capital city is Hanoi.

Virgin Islands

An island group that is divided between the U.S. and Britain. The Virgin Islands of the U.S. lie in the Caribbean. The capital city is Charlotte Amalie. The British Virgin Islands are in the Atlantic

Ocean a little further to the east. Road Town is the capital city. Both island groups are favorite tourist attractions because of the delightful tropical climate and scenic beaches.

Virginia

One of the thirteen original states in eastern U.S. The capital city is Richmond.

Vistula

A river in Poland that flows from the Carpathians north to the Gulf of Danzig.

Vladivostok

A city and port in southeastern Russia.

Volcano

A vent in the crust of the earth through which molten rock, called lava, pours forth. The lava originates in giant magma chambers that are subjected to intense pressure and heat. Also a mountain that results from eruptions. Over time and continued activity, ash and cinders grow into a mountain called a volcano. Volcanoes can be active (Vesuvius), dormant (Paracutin), or extinct (Killimanjaro). See also; Crater.

Volga

The longest river in Europe. It drains a huge basin in the Commonwealth of Independent Nations and empties into the Caspian Sea through a labyrinthine delta.

Vulcanicity

The action of a volcano.

Wabash

A tributary in Indiana and Illinois that flows into the Ohio river.

Wadi

A river valley that is dry except during the rainy season.

Waikiki

A popular resort beach in Honolulu, Hawaii.

Wales

A Celtic country on the western flanks of England. Wales is one of the four countries that make up the United Kingdom of Britain. The capital city is Cardiff. The Welsh landscape, to the south, is heavily industrialized because of the presence of coal mines. But the northern areas are remote and picturesque.

Warm front

The forward edge of an advancing mass of warm air, that displaces colder air.

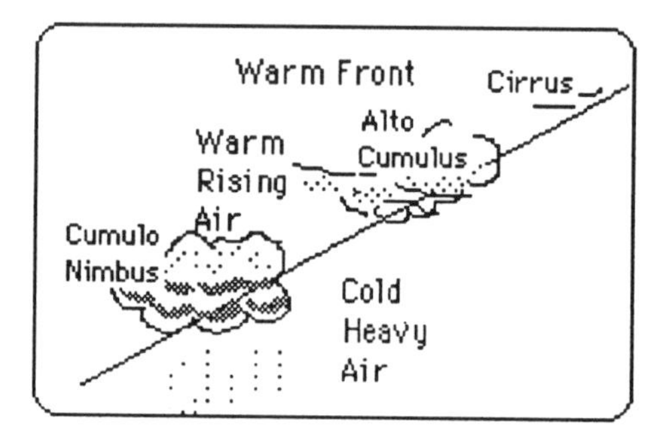

Fig. W i Warm Front

Warsaw

The capital city of Poland on the river Vistula.

Washington

A state in northwestern U.S. Olympia is its capital.

Fig. W ii Washington State

Washington D.C.

The capital city of the U.S. coexistent with the District of Columbia. It is situated on the Potomac river, between West Virginia and Maryland, in the northeastern portion of the country.

Waste disposal

The treatment method that is utilized by modern man for sewage and industrial effluent.

Water Cycle

The circulation of water in our ecosystem. First, it evaporates from the ocean into the air and becomes water vapor. There it cools, condenses, forms into clouds and falls back to earth as precipitation. From there, it seeps into the ground, runs-off into rivers or evaporates to make its way back to the ocean again. It is known also as the hydrological cycle.

Water management

The planned utilization of nature's water resources.

Water power

The energy that is endemic in running water, usually to benefit man by generating electricity.

Water table

The level below which the ground is saturated with water.

Water vapor (Atmosphere)

Water present in the air in the form of a gas.

Waterfall

A cascade of water over a fall like a rock barrier.

Waterfall, Hardanger Fjord Norway

Watershed

A ridge dividing the areas drained by different river systems. See also; Divide.

Wave

A motion that carries energy from one place to another.

Wave action

The movement of water in the ocean as it circles in a wheel-like motion, spurred on by the prevailing winds.

Wave-cut platform

A flat platform near the water-line that has been eroded by the wave action of the sea, especially in stormy conditions.

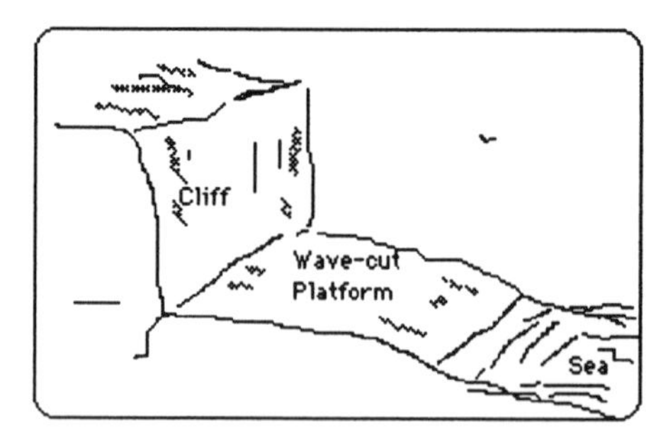

Fig. W ii Wave-cut Platform

Weather

The condition of the atmosphere at any given time.

Weathering

The mechanical and chemical breaking down and wearing-away of rocks by nature's elements (rain, ice, plant roots etc.)

Well

A spring where fresh water emerges from the bedrock.

Wellington

The capital city of New Zealand situated on the North Island on Cook Strait.

West Indies

A chain of islands that lies between the Caribbean and the Atlantic Oceans. The islands include the Bahamas, the Greater Antilles - Cuba, Jamaica, Puerto Rico and Hispaniola - and the Lesser Antilles.

Westerlies

The prevailing winds that flow from the west.

West Virginia

A state in the eastern U.S. whose capital is Charleston.

Western Australia

A state in west Australia on the Indian Ocean, whose capital is Perth. See also; Australia.

White Dwarf

A small but extremely dense star.

Whitney, Mount

The highest point in the U.S. outside of Alaska, at 14,494 ft. This mountain is situated in the Sierra Nevada range in Sequoia National Park, California.

Wind

Air moving across the surface of the earth. Air moves from areas of high atmospheric pressure to areas of lower atmospheric pressure.

Wind break

A line of trees that have been systematically planted to break the force of the wind, decrease erosion and lessenthe danger of fanning forest fires.

Wind - speed and direction

The task of an anemometer is to measure the velocity of the wind and indicate the direction from whence it originated.

Wind power

The force by which the wind blows. It can be put to use to drive engines, turbines and mills.

Wind vane

An instrument that is blown by the wind and is used to tell its direction.

Winnipeg

A Canadian city, the capital of Manitoba. Also a lake in Manitoba drained by the Nelson river.

Winter solstice

The time in the northern hemisphere when the sun shines directly down on its most southerly point, 23.5' below the equator, usually on December 21/22. See also; Solstice.

Wisconsin

An agricultural state in north central U.S. noted for dairy produce. The capital city is Madison.

World

Planet earth - our home.

World map

A map of the globe on flat paper.

World standard geographic coordinate system

The standard method for coordinating features and measurements that pertain to geographic phenomena.

Wyoming

A state in northwestern U.S. noted for its expansive grazing grounds. The capital city is Cheyenne.

X-rays

Radiation of extremely short wavelength, that can pass through solid objects. X-rays have many productive uses in medicine, commerce and science.

Xanthus

A river in Turkey that flows southwest into the Mediterranean. A city of the same name is sited at its mouth.

Xhosa

A black people, whose ancestors have inhabited South Africa since the sixteenth century. Traditionally herders, many Xhosa have moved to urban areas in search of work. These people have suffered much discrimination under the policy of segregation called apartheid.

Xingu

A river in Brazil that flows north from the Mato Grosso Plateau and confluences with the Amazon near its mouth.

Yangtze

The longest and most important river in China. It flows from the Tibetan highlands into the East China Sea.

Yardang

A desert ventifact caused by aeolian erosion.

Year

The period of the earth's revolution about the sun. It is 365 days, 6 hours and some minutes for our purposes of creating and living by a calender.

Yellowstone National Park

A famous park-land in northwestern Wyoming and environs, that contains spectacular geothermal features, including geysers, fumaroles, lava lakes and other dramatic scenery. The home of a famous geyser called *Old Faithful,* Yellowstone was the first national park to be established in the U.S. and the oldest in the world.

Yemen

A country in the southwest Arabian peninsula bordering on the Gulf of Aden. The capital city is Aden.

Yenisey

A river in Siberia, a province of Russia, flowing north into the Arctic Ocean.

Yosemite

A famous park-land in California that contains spectacular glacial features, including U-shaped valleys, hanging valleys, waterfalls and other dramatic scenery. Yosemite Falls is one of the highest falls in North America and attracts millions of tourists annually.

Yukatan

A large peninsula that separates the Gulf of Mexico from the Caribbean. It includes parts of Mexico, Guatemala and Belize. Ruins of an ancient Mayan civilization have been discovered in this location.

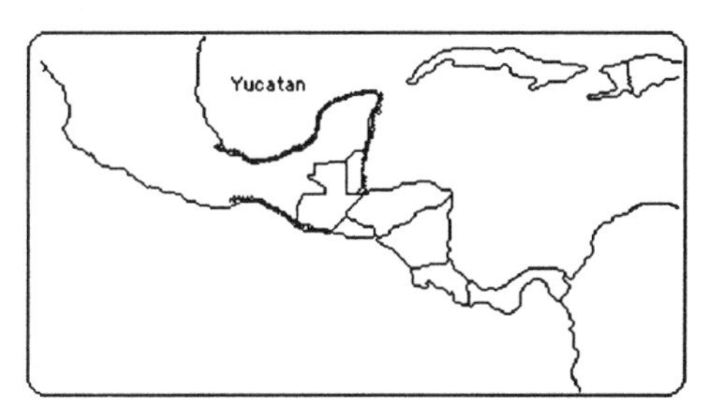

Fig. Y i Yukatan

Yukon

A rugged territory in NW Canada between Alaska and British Columbia.

Yukon River

One of the longest rivers in North America. It flows through the Yukon Territory and Alaska and empties into the Bering Sea.

Yurt

A circular framed tent used by the nomads of Mongolia.

Zaire

A tropical country in central Africa on the equator. It has one of the world's most lush tropical forests. The capital city is Kinshasa.

Zambezi

A river flowing in southwestern Africa into the Mozambique channel.

Zambia

A country in south central Africa, south of Zaire. Its economy is based on the production of copper and tropical exports. The capital city is Lusaka.

Zealand

Largest island of Denmark in northern Europe.

Zealand (Netherlands)

A region in southern Netherlands, where huge artificial dikes keep the sea at bay.

Zeugen

A desert landscape feature as a result of aeolian erosion.

Zimbabwe

A country in southern Africa, formerly known as Rhodesia. The capital city is Harare.

Zion National Park

Beautiful park-land in Utah noted for its spectacularly sculptured rock formations.

Fig. Z i Zion

Zone

One of the five great latitudinal divisions of the earth's surface named according to the prevailing climate. There are two frigid zones surrounding the polar regions and one torrid zone between the tropics of Cancer and Capricorn. The two temperate zones correspond to the mid-latitude regions north and south of the equator.

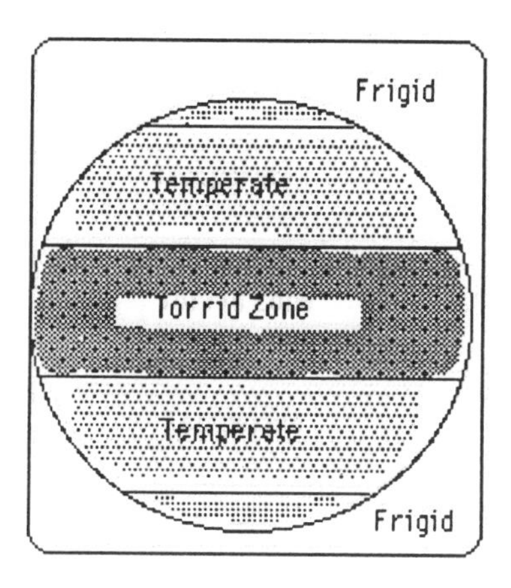

Fig. Z ii Zones

Zuider Zee

A section of the North Sea that extended into the Netherlands and was cut off and dammed by dikes to make a large man-made lake, called Iselmeer. See also; Netherlands.

Zulu

A pastoral people, most of whom live in Natal, a province in South Africa. Zulu people were subjected to severe discrimination as a result of the South African policy of apartheid.